Jack Ashby is the assistant director of the University Museum of Zoology, Cambridge, and an honorary research fellow in the Department of Science and Technology Studies at University College London. He is the author of *Animal Kingdom: A Natural History in 100 Objects* and lives in Hertfordshire.

Winner of the Whitley Award
for Best Natural History Book 2022

'Ashby reveals marvellous creatures and the mysteries and myths surrounding them'   BBC *Wildlife Magazine*

'Charming, informative . . . Ashby's intoxication with Australia's mammals makes for a marvellous read . . . *Platypus Matters* is full of astonishing facts that are certain to have you thinking differently about Australia's unique mammalian fauna.'
Tim Flannery, *New York Review of Books*

'Hopefully, Ashby's book of wonders will not only delight, but will also inspire increased efforts to preserve Australia's wildlife through the perilous decades to come.'
Natural History

'Building on his considerable scientific knowledge and decades of field experience, Ashby immerses readers in all things platypus . . . *Platypus Matters* is a must-read for any mammal nerd or Aussie wildlife enthusiast.'
Justine E. Hausheer, Nature Conservancy's *Cool Green Science*

'An engaging natural (and enraging colonial) history . . . Given our seemingly unwavering commitment to destruction via global warming and environmental predation, Ashby has his work cut out for him in making us care about the natural world's most unusual inhabitants. Still, his efforts, like those of everyone pushing back against what feels like an unstoppable tide, are crucial to building a coalition that understands the platypus – like the planet sustaining it and us – matters.'

Washington Independent Review of Books

'From platypuses and possums, through wombats, echidnas, devils and kangaroos, to quolls, dibblers, dunnarts and kowaris, Ashby knows them all; and in his recently published *Platypus Matters*, he guides his readers on a tour of their lives, their evolutionary stories and the challenges they face in the modern world.'   *The Well-read Naturalist*

'Fascinating . . . This is wonderfully dorky stuff . . . Ashby energetically recounts the stories of a dozen species of Australian animals, their natural histories, and the universally-tragic record of their discovery by, interaction with, and decimation or extermination at the hands of humans. These are largely depressing stories, a long Australian list of ecological travesties that only look more dire when read against the backdrop of the environmental catastrophes stalking the country in the twenty-first century. And although *Platypus Matters* is a persistently, defiantly upbeat book, downright infused with Ashby's scientific exuberance, it can't avoid a touch of the funereal. In Australia as everywhere else, the ecological outlook for all forms of life humans don't enslave, exploit or eat is bleak . . . But considering the wonders at stake, any start is a good start.'

Open Letters Review

'Keen to overturn the warped, colonial perception that monotremes (e.g. platypuses and echidnas) and marsupials are more primitive than other mammal species, the zoologist author . . . takes us on a tour of the fauna of Australia in all their glory. In an engaging and entertaining narrative reminiscent of Gerald Durrell, we learn that wombats produce cubic poo, that the platypus played a disruptive role in the narrative of evolution, and much more besides.'

'This is a compelling, funny, firsthand account of our wonderfully unique mammals and how our perceptions of them impact their future.'

'Ashby has an infectious enthusiasm for Aussie marsupials and monotremes'

'Written in a lively, conversational style and drawing on decades of fieldwork, this is a beguiling portrait of our unique fauna.'

'*Platypus Matters* contains a wealth of information on the natural history and biology of Australian mammals . . . The author weaves together a personal story about his experiences in Australia and his fascination with its fauna with a wealth of biological facts and strategic historical anecdotes.'

'Ashby makes the case that Australia's wildlife is not a collection of oddities and species that can kill you, as it is most often, and even well-meaningly portrayed. He explores how this traditional narrative about Australia's native animals arose, how it is incorrect, and shows why it matters. Some of the species met along the way, including echidnas, wombats, Tasmanian devils and scaly-tailed possums, leave lovely impressions that will be lasting portrayals. Both serious and fun, *Platypus Matters* is compelling reading.'

'*Platypus Matters* is an original, charming book with a contemporary message. Ashby seeks to convince us of the importance of Australia's mammals, using the platypus as a worthy ambassador. Most importantly, with a combination of beguiling stories and impassioned arguments, he explains the very real consequences of devaluing Australian wildlife for the survival of this unique fauna. Ashby's raw enthusiasm as a naturalist and love of sharing a good anecdote make for entertaining reading, but the final chapters take a sobering turn, and rightly so. His book is a clear call to action to address the urgency of the current extinction crisis. It's also a bloody good read.'

Katherine Tuft, general manager, Arid Recovery (Australia)

'Timely, important and multifaceted, *Platypus Matters* is a lesson in the evolution of mammals, a historical journey and an adventure book packed with exciting stories of Ashby's global travels. Most profound is the book's intellectual exploration of colonial perspectives and how they shaped the world's understanding of and subsequent relationship with Australia's unique fauna – to this day. Fascinating and enlightening. Only Ashby could have written this book, and I absolutely loved it!'

Georgia Ward-Fear, Macquarie University,
cofounder, Cane Toad Coalition

JACK ASHBY

Platypus Matters

*The Extraordinary Story of
Australian Mammals*

The University of Chicago Press

The University of Chicago Press, Chicago 60637
© 2022 by Jack Ashby
All rights reserved. No part of this book may be used or reproduced in any
manner whatsoever without written permission, except in the case of brief
quotations in critical articles and reviews. For more information, contact the
University of Chicago Press, 1427 E. 60th St., Chicago, IL 60637.
Published 2022
Paperback edition 2024
Printed in the United States of America

33 32 31 30 29 28 27 26 25 24   1 2 3 4 5

ISBN-13: 978-0-226-78925-5 (cloth)
ISBN-13: 978-0-226-83321-7 (paper)
ISBN-13: 978-0-226-78939-2 (e-book)
DOI: https://doi.org/10.7208/chicago/9780226789392.001.0001

Published in Great Britain by William Collins, an imprint of
HarperCollins Publishers Ltd., 2022.

Library of Congress Cataloging-in-Publication Data

Names: Ashby, Jack, author.
Title: Platypus matters : the extraordinary story of Australian
    mammals / Jack Ashby.
Description: Chicago : University of Chicago Press, 2022. | Includes
    bibliographical references and index.
Identifiers: LCCN 2021036608 | ISBN 9780226789255 (cloth) | ISBN
    9780226789392 (ebook)
Subjects: LCSH: Platypus. | Mammals—Australia.
Classification: LCC QL737.M72 A84 2022 | DDC 599.2/9—dc23
LC record available at https://lccn.loc.gov/2021036608

♾ This paper meets the requirements of ANSI/NISO Z39.48-1992
(Permanence of Paper).

*To Australia's naturalists,
past, present and future.*

Contents

Preface

Ever since I first encountered them as museum specimens at university, platypuses have been my favourite animals. It may seem childish that a grown adult – particularly one that works in science – has a favourite animal, and perhaps it is, but the more I learn about them, the more I am convinced that nothing more wonderful has ever evolved.

Following those undergraduate classes, finding platypuses in the wild shot straight to the top of my zoological to-do list. All zoologists have these biological bucket lists and they typically define how nature-nerds spend their time, working hard to find and observe the animals that fascinate them most.

As well as platypuses, my list also includes species that are relatively widespread and found closer to home. For example, over the years I have spent what adds up to several fruitless weeks sat at the edge of countless British bodies of water failing to see a Eurasian otter (the closest thing we have to a platypus, ecologically speaking), until I finally found one in a stream running through the middle of the town of Frome in Somerset, while teenagers loudly performed donuts in the supermarket parking lot right alongside.

It sometimes feels as though these hard-earned encounters divide one's life into discrete chunks. 'Before you were stared down by a family of snow leopards' and 'after you were stared down by a family of snow leopards'. At other times, though, these moments can be deeply frustrating: I only know that I have been in the presence of a sloth bear from seeing the two reflective dots of its eyes as it noisily snuffled for fallen fruit in the pitch black beyond the limits of my torch-light. To have continued closer would have been foolish with such an unpredictable and well-clawed animal.

Less dangerous, my first wombat sighting caused me to start shrieking excitedly at the driver on a packed bus as we sped past the animal I had dreamed about seeing for years. She didn't stop the bus, and I was heartbroken that the encounter was so fleeting. Until, that was, we got off at the next stop and found perhaps fifty wombats wombling around the immediate area. Each moment with a sought-after species becomes burnt into the memory like other significant life events.

For me, the idea of a zoological bucket list isn't about ticking boxes on an animal bingo card. Working with museum specimens, reading descriptions or watching footage can only go so far in furthering our appreciation for a species. We can never hope to truly *know* an animal, only to try, but nothing beats seeing them – being there with them – on their own terms and in their own environment as they go about their business. An exercise in list-ticking would involve being satisfied by just a single encounter, before moving on to the next species down the list. Instead, finding a 'passion species' typically means you want to keep working to see them again.

A year into my first proper job at the Grant Museum of Zoology in London, I had saved enough money for a return flight to Tasmania to search for the first animal to make my list: the platypus.

In our first few days in Tasmania, my friend Toby Nowlan (now a wildlife filmmaker) and I managed to find wombats, Tasmanian devils, quolls (slender, spotted mongoose-like relatives of the Tasmanian devil) and echidnas (platypuses' spiny ant-eating relatives), but had not set eyes on a platypus. We were walking the Overland Track across the Tasmanian highlands – one of Australia's great long-distance treks – at the height of summer. This week-long hike takes you through temperate rainforest, buttongrass moors and mountain passes, and via many lakes and streams that are perfect for spotting platypuses.

Unfortunately, the Overland Track in summer is so beautiful that it is also extremely busy. Every evening we would head for the nearest body of water and look for platypuses, only to find that other walkers had made it there first for a relaxing swim, which reduced our chances of success to zero.

Tasmania is famous for its ability to swing rapidly from T-shirt weather to freezing conditions, even in midsummer. On the penultimate leg of the trek – on the day before Christmas Eve – the weather broke for the worse, and it started to snow. The day's route took us down out of the mountains, descending to the northern tip of leeawulenna, or Lake St Clair, Australia's deepest lake. With the drop in altitude, the snow turned into rain. Torrential, rainforest rain. We were thrilled: with the other walkers sheltering safely in the nearest hut, Toby and I waited for evening and headed to the lake.

As mammals, foraging platypuses must return to the surface to breathe, giving platypus-watchers a few seconds of precious viewing – interspersed with what feel like interminable periods when you fear that they have disappeared into their burrows. Such brief appearances are probably for the best, though, if – like me – you involuntarily stop breathing while they are in sight. Holding one's breath could be wise, however, as platypuses are extraordinarily sensitive to both movement and sound when at the surface. The tactic for approaching them is like a children's party game: you creep closer, but if the platypus sees you move, you're out.

Platypuses are small, so the challenge is to get close enough to get a good look. While they are underwater, you can move and communicate with your fellow watchers as much as you like, because the platypuses' eyes and ears are closed. But when the platypuses are on the surface you must become a silent statue, or they will disappear. Game over.

Most platypus dives last less than a minute, and Toby and I had agreed our strategy. If either of us spotted one, we would wait for it to dive, and then give ourselves a maximum of thirty seconds to get the other person's attention while also creeping closer. After that time, we'd have to freeze, remain silent and wait for the platypus to reappear, then repeat the process each time it dived until we were at the nearest point on the shore to the animal.

The payoff for hours of platypus-watching is often just a series of very brief glimpses (they dive around seventy-five times each hour when foraging). As frustrating as that can be, the avid platypus-spotter does also have to recognise that, metabolically, it is very impressive. If I spent a minute underwater, I would certainly need more than ten seconds at the surface before going under for another minute – and I wouldn't be chewing my dinner in that same, short window. Platypuses have a number of behavioural and physiological adaptations to allow for this: their blood has a high oxygen-carrying capacity; their lungs are large so they can take in a lot of air at once; and whereas their heart can beat at 230 beats per minute (bpm) at the surface of the water, they can reduce this to as slow as ten bpm while diving, which is a dramatic reduction.[1]

And so there we were, on the bank of a river, close to where it joined the massive lake. I had found the entrance to a burrow near the water's edge and decided to sit by it and wait. Toby had wandered on downstream towards the lake. An hour passed and I realised that we were no longer in earshot of one another. I know this was selfish, but I thought the one thing worse than not seeing a platypus would be if one of us saw one but the other didn't. I got up and followed Toby's tracks.

Through the rain, I heard him calling. I ran. He was on

the other side of a marsh by the lake. I could tell by the look in his eyes that he had seen a platypus. In him, joy. In me, terror. What if it didn't come back? How would our friendship survive that?? Toby pointed – and then it happened. In the freezing, sodden, dying light of 23 December 2005, I saw my first platypus as it surfaced about 30 metres (100 feet) away.

We followed the rules. We froze. We didn't utter a sound. It's likely I had tears in my eyes, but in the rain who could tell? We watched. It dived. We ran. It surfaced. We froze. We watched. It dived. We ran. On the third time, it surfaced 1.5 metres (5 feet) away, right in the shallows. We could make out the lighter spots next to its eyes and judge the suppleness of its bill. It dived again. I ran. I moved for my allotted thirty seconds then stopped and waited once more.

For reasons unknown to humankind, the UK has developed the slapstick theatrical tradition of pantomime. One of the many integral audience-engagement tropes of pantomime is for one of the cast to engage in a back-and-forth argument with the audience about whether or not another character has just done something on stage. They say something like 'Oh yes he did!' to which the audience chants back, 'Oh no he didn't!' This is repeated three or four times until at last the cast member reverses tack, proclaiming 'Oh no he didn't!' and thereby tricking the audience into agreeing with his or her original point when they respond with the opposite: 'Oh yes he did!' (Apologies to anyone who has never been to a panto, but this is typically the high point of the show. It is exactly as banal as it sounds.)

Anyway, this is how I would describe what happened on the lakeshore that night. The platypus and Toby and I had been engaging in a back-and-forth of surfacing/freezing and diving/running. On the fourth occasion it dived, I ran and then froze – but it didn't reappear when I expected it to. Perhaps a further minute passed. In order to hide from danger, platypuses can wedge themselves under roots or branches underwater and dramatically decrease their metabolism to remain submerged for ten minutes or more. I remained frozen. When it eventually reappeared, to my mind it was as if it had reversed the pantomime 'argument'. I got the rules muddled and ran. Inevitably it disappeared with a splash, for good. I am still cross with myself.

Happily, I have seen many platypuses since then, and have re-walked that same Overland Track on other occasions, even at the height of winter when I've found myself alone on the snow-covered route for the entire week. With no human swimmers to disturb them, platypuses can then be found in every lake along the way. The water partially freezes over, but platypuses do not hibernate throughout the winter and as such must surface in the ice-free patches. As they can only surface in a limited area of free water, spotting them is easier, but sitting through the freezing winter evenings, often with rain, snow and strong winds, can be fairly uncomfortable. Obviously, I think it's worth it.

*

This book is about more than the platypus. For me, that species acted as a gateway to the marvels of all Australian mammals. Similarly, through the pages that follow, I start with platypuses, but go on to explore the other mammals with which they share their country – echidnas, marsupials, bats and rodents. The story will be influenced by ecological fieldwork, the specimens we find in our museums and historical views of these astonishing animals.

Australasia has it all in terms of diversity of mammal reproduction. It is the only place on Earth where you can find all three ways of being a mammal: monotremes, marsupials and placentals. Monotremes are mammals that lay eggs (today, represented by the platypus and echidnas); marsupials are mammals that give birth to tiny live young after a short pregnancy, with the young doing most of their infant growth during a long period suckling milk, often in a pouch; and placentals are mammals that give birth to young at a later stage of development after a long pregnancy, the young finishing off their infant growth by suckling milk for a relatively short period (we are placental mammals, as are most mammals, including rodents, cats, antelopes and whales).

I have come to notice that these three groups are often not considered equals by the world at large, and that's why I wanted to write this book. Australian mammals are looked upon fondly, but not fairly. It is often implied that monotremes and marsupials are somehow lesser than other mammals. Worse – that they are *biologically determined* to be lesser. I'm going to show you the subtle (and not so subtle) ways that this can take shape, and try to explain why such a hierarchy has been politically and philosophically fabricated, perhaps subconsciously.

Most importantly, I'm going to argue why it matters that we think about how these animals are portrayed – how we talk about them, how we represent them on TV and in museums, and how we value them. Australia's unique wildlife is disappearing at a rate unparalleled by any other large region on Earth, and its conservation is surely tied to how these animals are understood. The way I see it, the fates of both the people and animals that live in Australia have in large part been determined by the way its wildlife has been presented in the West. And that story begins when Europe met Australia's mammals.

Introduction

In December 1799, a woman was walking from the dockyard to the Assembly Rooms in Newcastle upon Tyne, in the northeast of England, making for the rooms that the town's Literary and Philosophical Society had rented to house its library and growing specimen collection. On her head she balanced a barrel of alcohol. According to an 1827 account, as she arrived, this barrel dramatically burst, spilling its precious contents over her in a cascade of 'pungent and foul-smelling spirits', when she was 'almost suffocated, if not drowned'.[1] In the process, this woman, whose name was not recorded, became one of the first Europeans to touch a platypus: it hit her on the head.

This momentous platypus shared its barrel with Europe's first wombat (adult wombats weigh nearly 30 kilograms/70 pounds, so presumably it presented this poor woman with the greater risk of head injury). Travelling halfway around the world together, the animals had been sent to the society by one of its more eminent members, John Hunter, the second governor of the new British colony of New South Wales in Australia. Hunter had originally kept the wombat alive to 'observe its Motions', but he had failed to work out what it ate, so it died of starvation after six weeks. In the barrel, he had managed to preserve all but its eyes, brain and bowels.[2]

Regarding the platypus, Hunter apologised that, 'the Weather having been exceedingly Warm when the Animal was killed, it could not be kept until we could have an opportunity of preserving it in Spirits,' so he only included the skin. The barrel was sent from Sydney to London, from where he asked Sir Joseph Banks, arguably then the world's most influential scientist, to forward it on to Newcastle.[3] Banks wasn't in town when the vessel containing the platypus and wombat arrived in London, so the society had to wait for months to get their hands on it.[4] In these days of colonial science, communications – and specimens – were often sent via important 'handlers' back home. If your handler wasn't available, however, correspondents could be left hanging.

*

As with so much information about Australia, Europe is indebted to Indigenous knowledge for its earliest encounters with the platypus. In November 1797, on a lagoon near the Hawkesbury River, Hunter had observed a Darug man (the Darug are a group of Indigenous Australians from the area that now incorporates Sydney) watching 'an amphibious animal of the mole kind' surface and dive for over an hour, until the man had the opportunity to spear it in the neck and forelimb.[5] It is likely that this was the specimen sent to Newcastle. Nevertheless, Hunter has been given the credit for this early European specimen.

Unquestionably, the most comprehensive publication on platypuses is Harry Burrell's exquisite 1927 book, *The Platypus*. Burrell's contributions to platypusology were groundbreaking, and I will quote him often in this book. He observed countless platypuses in the wild and in elaborately staged experimental enclosures (the details of his experiments will make you wince – they would not pass ethical scrutiny today), and he collected and dissected them to determine much of what we now know about the species. However, he also dismisses Aboriginal

The fubjoined ENGRAVING is from a DRAWING made on the fpot by GOVERNOR HUNTER.

ORNITHORHYNCUS PARADOXUS.

AN AMPHIBIOUS ANIMAL of the MOLE KIND,

which Inhabits the Banks of the fresh water Lagoons in New South Wales - ?
its fore feet are evidently their principal affistance in Swimming & their hind
feet having the Claws extending beyond the Webil part are useful in burrowing.

'An amphibious animal of the mole kind' – John Hunter's own drawing of the platypus specimen he acquired from a Darug man in 1797, reproduced in David Collins' 1802 book, *An account of the English colony in New South Wales.*[6]

Australians' contributions to scientific knowledge when he describes Hunter's account of watching the Darug man procure this historic specimen. He precedes a full report of the event with the phrase 'there occurs the following account of what was apparently the first platypus captured by a European.'[7] But Hunter didn't capture the specimen; the unnamed Aboriginal man did.

For Europe, the discovery and description of the platypus – and the echidna a few years earlier – was more than just the addition of a new species to the catalogue of known life. This key moment in the northern hemisphere's relationship with Australia's wildlife (and, I argue, the relationship with Australia as a whole) caused the leading scientists of the time to reconsider how some of life's major animal groupings were arranged and interrelated. The biology of the platypus and echidna became a touchstone for the battle around evolution. Quite unlike anything anyone in Europe had seen before, the creatures possessed a mysterious mix of features which science had previously characterised as *either* mammalian, reptilian or avian. No animal should have been in possession of qualities seen in more than one of these groups, yet the platypus and echidna did. And so, the question became: where should scientists place them in the tree of life?

It wasn't until 1884, after nearly a century of controversy, involving some of the biggest names in the history of science and an extraordinary number of slaughtered animals, that scientists were finally satisfied that platypuses and echidnas truly are mammals that lay eggs.

Today, this epic nineteenth-century dispute continues to colour the way the world sees these species. Their divergence from the way that Western naturalists then arranged the animal kingdom means that they are still too-often characterised as primitive or weird. And platypuses and echidnas are not alone in this. All of Australia's native fauna – particularly the marsupials but also the placentals – has been affected.

Think about it for a moment. How would you describe Australian animals? As someone who has chosen to read this book, you may well think they are as wonderful as I do. But even so, many otherwise enthusiastic testimonials for the greatness of Aussie animals tend to include words like weird, strange, bizarre, dopey or even primitive. And this is often followed by a phrase along the lines of 'everything there is trying to kill you'.

These are, after all, the typical ways in which Australian wildlife is described in documentaries, media articles and even natural history museum galleries. It's fair to say that the words 'weird', 'bizarre' and 'strange' are all intended to be playful and to conjure mystery. And the idea of a country where everything from trees to snails is venomous is probably supposed to evoke a sense of romantic adventure. However, each of these descriptions also has negative connotations. And no species is more a victim of this implicit disdain than the platypus.

Platypuses and echidnas are said to share certain important, noticeable features with reptiles, such as egg-laying and the way they walk,* and this has led to them regularly being described as primitive. This is unfair. I experience a deep shudder of injustice when my work takes me to a museum that incorporates the platypus and its relatives in a display of 'primitive mammals' (the US is particularly guilty of this). Throughout scientific literature, the notion that platypuses are primitive is casually dropped into introductory statements. I came across a typical example in an academic book about the blood-clotting adaptations of different groups of animals – platypuses featured (along with wallabies, to which they are only distantly related) in a chapter titled 'Primitive Australian Mammals', without any justification for this hierarchical view of nature.[8] Platypuses are not primitive. Nor indeed is any other living species.

*

My job is as the assistant director of the University Museum of Zoology in Cambridge, which holds one of the UK's largest and most significant natural history collections. However, each year I leave behind our two million dead animals and head to Australia to undertake ecological fieldwork for Australian charities and universities, surveying mammals that are very much alive. Through both parts of my professional life – working with the dead and the living – I have come to the conclusion

---

* It is a widespread notion that both monotremes and living reptiles walk with a sprawling, side-to-side gait – a trait they are both said to share with the reptile-like group from which mammals evolved. However, studies of all three groups have called into question the extent to which both living monotremes and their extinct ancestors do walk like lizards. (Pridmore, 1985; Jones et al., 2021)

that Australia has the most incredible, interesting and *wonderful* wildlife on Earth. I've followed animals across Himalayan snowfields, African plains, Arctic tundra, Indian jungles, European taiga forests and American prairies. But, for me, Australia should be *the* ecological mecca for any eager mammal-watcher. As I hope to convey through the course of this book, no other place can hold a candle to its mammalian wonders.

This is not, however, the prevailing view. There remains an unscientific, unjustifiable prejudice against Australia's marsupials and monotremes. This manifests in many ways, but one example is that Australian mammals were assumed to be less intelligent and have smaller brains than other mammals (comparing an animal's brain size to its body size has often been used as an indicator of intelligence), and this is popularly given as a key reason for their high modern extinction rate. Their portrayal as stupid animals was established early on. For example, Victorian England's most eminent naturalist, Richard Owen, claimed that the various marsupials he had encountered at London Zoo 'are all characterized by a low degree of intelligence'.[9]

It turns out, however, that over the course of nearly 250 years of marsupial biology in Europe, scientists had been regularly repeating this as fact without anyone having actually checked to see if it were true. It isn't. In 2010, my colleagues Vera Weisbecker and Anjali Goswami calculated the brain sizes for a wide range of marsupial and placental mammals. They found that the assumption that placentals are brainier than marsupials was being skewed by one particular placental group, the members of which have extraordinarily large brains: the primates. They discovered that if primates were taken out of the equation, there is no difference: marsupials do not have smaller brains than placentals of similar body size. In fact, at smaller body sizes, marsupials have relatively larger brains.[10] This is not a difficult calculation to perform. It didn't even require access to brain specimens (which might have been hard to come by, certainly in the past): Anjali and Vera determined the animals' brain sizes by pouring tiny beads into skulls held by museums, and then measuring the volume of those beads. It is remarkable that it took until 2010 for anyone to think to do this.

This kind of anti-Australian bias seems to be based on the mystifying notion that placentals – mammals like us, found across the rest of the world – are inherently superior. This is just our egos talking.

Take, for example, naturalist Harry Burrell. In his seminal book, he took a clear hierarchical view of the animal kingdom. He begins by saying that the platypus is 'a primitive kind of mammal, which is in some respects intermediate between the higher mammals and the reptiles'.[11] This single sentence includes the notion that platypuses are primitive *and* the idea that some mammals are 'higher' than others on a great chain of being. He wasn't alone in the use of such words, and indeed it remains commonplace today. For example, how often have you heard the phrase 'great apes' used to refer to chimps, humans, gorillas and orangutans, in order to distinguish them from gibbons, the notional 'lesser apes'? Wouldn't 'large apes' and 'small apes' be less of a value judgement? They are both equally evolved.

This gothic description of the platypus by the eminent physician William Kitchen Parker in his 1885 lectures to the Royal College of Surgeons is similarly reductive:

> Here is a beast – a primary kind of beast, a Prototherian – whose general structure puts it somewhere on the same level as low reptiles, and old sorts of birds . . . they are all equally below the morphological level of the nobler Mammalia.[12]

All language like this is inherited from the thoroughly outdated view that evolution is progressive. It isn't. When we lend credit to the impression that some animals are more or less 'advanced' than others, we imply that evolution works like a ladder travelling up the tree of life. The tree analogy is useful in that life only has a single trunk – it only evolved once – and over time taxonomic groups split from one another and became branches that further split over time – shaped in part by their neighbours, but separate from them – with species like leaves at the branches' ends. But the tree model implies altitude is gained over time as it grows, which isn't helpful. I prefer to think of evolution as a flat, forking path with no known destination. As we travel forward in time, we can trace a group's evolutionary history and see its relationships by taking paces *along* the branches that form in the path, rather than *up* them.

The tree model is further problematic when humans, or other alleged 'higher mammals', are placed at the top, as if we are the pinnacle – the deliberate end point of around four billion years of evolution. We are just one leaf at the end of one branch.

Arguably, one of the most influential popularisers of natural history in Victorian Britain was John Gould. In the introduction to his 1863 work *The Mammals of Australia*, he supposes that the species there are stuck in a lowlier form of development: 'I may ask, has creation been arrested in this strange land?'[13]

Such negative descriptions do not stand up to scientific scrutiny. Why are we so mean about Australian wildlife? I believe it reflects a continuation of a prejudicial colonial mindset: an unconscious assumption that life in the 'colonies' is fundamentally second-rate compared to the noble northern hemisphere. While this treatment is widespread, it stems from a preconception to which most people would deny that they subscribe. It is unintentional but almost universal.

When I discussed this with curator and historian Subhadra Das, who is researching the history of eugenics, she pointed out that all this talk of brain sizes is strongly reminiscent of the anthropometric laboratories that sprung up across the West towards the end of the nineteenth century and into the twentieth century. At the time of our conversation, Subhadra was responsible for curating the Galton Collection at University College London. Francis Galton established the racist pseudo-science of eugenics. Anthropometric laboratories like his and others elsewhere in Europe and North America sought to determine anatomical measurements that could be used to demonstrate that different groups of humans had heritable biological features that could be systematically classified. One 'result' was the false claim that black people had smaller brains than white Europeans. This was used to justify a hierarchy among human races. Eugenicists used their 'science' to establish a notion that racialised people were destined – as a fundamental function of their biology – to be socially inferior to white people. In similar ways and at a similar time, incorrect arguments about brain sizes were being used to establish a false hierarchy among different groups of mammals. The arguments relating to humans and to other animals – I think – are both deeply grounded in imperialist goals.

Now, I need to acknowledge something that there is no getting around: I am a white European writing about and interpreting Australian animals. This itself could be considered an echo of colonialism. I have come to regard Australian mammals with deep awe and admiration, and first and foremost I want to share that with you. From there, this book goes on to explore why it's important that we consider whether colonial hierarchical narratives remain in the ways we talk about Australian wildlife. There is an undeniable tension in someone like me doing that. For any Aboriginal or Torres Strait Islander people reading this, I apologise that at times I am telling parts of your story back to you. I only hope that my experience of museum-based and field-based science, along with the exposure to the history of natural history that comes from working in a museum, can make a useful contribution to conversations about colonial structures that are taking place around the world.

*

The underlying prejudices about Australian animals have real global impact, specifically in two profound ways. First, I would go so far as to say that the concept of *terra nullius* – literally 'nobody's land', the suggestion that when Britain claimed Australia for the Crown the land didn't 'belong' to anyone, and certainly not its Indigenous peoples – was and is still in part supported by Western attitudes to Australia's native fauna as much as to its native people. This idea helped to legitimise the European invasion of Australia and the historical and ongoing displacement of its Indigenous peoples as well as its native wildlife. Platypuses are political.

Second, Australia has the worst recent mammal extinction record of anywhere in the world, and those species that survive are in crisis. The platypus and its neighbours really do matter: the way we speak about Australia's animals has consequences for how we value them, and therefore for the likelihood that we will protect them.

The catastrophic bush fires of the 2019/20 fire season appear to have heralded a new normal for Australia's climate. Early in the crisis, we saw reports of millions of animals dying as a result of the fires, and then it was half a billion, and then it was well over a billion. *Over a billion* animals died. It is hard to comprehend what death on that scale looks like, but by then it was clear that Australia's ecosystems could not afford to lose this volume of wildlife. The attitudes I'm talking about, at least in part, had led to a situation where Australia's unique fauna was not being valued by Australia's own government. Before the fires, this had already left many species in extremely precarious positions, with massively reduced ranges and vastly diminished population sizes. The International

Union for Conservation of Nature (IUCN), which assesses threats to species globally, had found that the populations of 102 species of Australia's native land mammals were shrinking, while only three were increasing (one of which – the burrowing bettong – or boodie – a small hopping marsupial – now only survives in fenced enclosures and on offshore islands). The IUCN considered the populations of 105 species as stable. It's worth pointing out that none of these assessments look more than a couple of decades into the past, so a 'stable' population (or increasing ones, like the bettong), could simply be one that dramatically crashed in the nineteenth or early twentieth centuries, but was no longer in decline at the time of the report. We don't even know whether a further seventy-three species were increasing, stable or decreasing.[14] And this was before the fires.

At the time of writing, we don't yet know how tragic the fires were, but in the immediate aftermath, conservationists were prevented from monitoring sites due to lockdown measures resulting from the 2020 coronavirus pandemic. This has hampered their ability to measure the true impact, but for many populations, it will have been the death blow. I'm sorry that this summary is so grim (such is the way of Australian ecology), but it explains why we should care about the way we talk about Aussie mammals – because there is still a lot to lose. And that includes the world's greatest animal – the platypus.

1

Meet the Platypus

Truly, platypuses are things of pure wonder. They combine an incredible array of characteristics and behaviours from across the animal kingdom as well as demonstrating qualities that are all but unique among mammals. The platypus has a body like a mole, feet like an otter, hamster-like cheek pouches and horny grinding pads in its mouth instead of teeth, which it uses to mush up insect larvae, crustaceans, worms and occasionally tadpoles and snails. And, most conspicuously, it has a beak, which at first appears almost indistinguishable from that of a duck – and this is why the earliest European settlers in New South Wales called them 'duck-moles' or 'water moles'.

As if all that weren't astounding enough, these semi-aquatic burrowing mammals lay eggs, but also produce milk – despite having no nipples (an old joke is that platypuses are the only animals that can produce their own custard). They hunt underwater with their eyes closed while sensors in their bills detect the electricity given off by their prey's beating hearts. Their fur fluoresces under UV light, but we don't know why (if indeed there is a reason).[1] And the males are venomous. They amaze me.

Since that first trip to Tasmania, I am lucky enough to have met platypuses regularly in Australian creeks and lakes, located in environments ranging from tropical rainforest to alpine tarns. Each instance is etched in my memory with undiluted joy. Combining these encounters on the platypuses' own terms with closer, unhurried meetings with museum specimens is a way of piecing together all the details of how they live their lives; to see what it means to be a platypus.

*

It's obvious to the most casual observer that platypus bills are unique: no other mammals have mouths anything like them. Nor does any other animal. The first Westerners to describe the platypus didn't immediately appreciate that their bills are not actually duck-like as they were working with dried skins sent back to Europe. The bills they encountered were tough and hard, whereas a live platypus's bill is supple and leathery in texture. Within the bill are thin rods of bone, which act as supports. Looking at their skulls in a museum, I always think that they resemble the tail of a massive earwig, as these rods look like pincers. But the skin that covers them is soft and rubbery.

Platypuses' bills are equipped with not one, but two sensory systems, which enable them to find food when the other senses have been shut off. Like most other aquatic mammals, platypuses close their nostrils when diving, eliminating smell (however, they do have an olfactory sense organ that opens into their mouth, called the vomeronasal organ, which could possibly be used underwater to detect chemicals given off by prey or the scent glands of other platypuses). In addition, their eyes and ears are positioned in a muscular groove in their skin, which they shut tightly while submerged, so they cannot use sight or sound to navigate or hunt either. (No other mammals pair their ears and eyes in this way – not even their relatives the echidnas.)

Instead, their bills contain mechanosensory cells, which means they are highly sensitive to touch. As is the case with many species of fishes, this may enable them to detect movements in the water that result from animals moving nearby. Their bills also have sensory mucous glands that can detect electric fields. As we are taught in biology classes, every single muscle contraction in the animal kingdom – including every heartbeat – is the result of an electrical signal from the nervous system. When hunting, platypuses scan their heads from side to side, moving them like a metal detectorist. We assume that they are tracking down live food by sensing the electricity being given off by other creatures' muscles. Sharks and some other fishes do this, too, but this behaviour is almost unheard of in mammals. To date, the only mammals known to be electro-sensitive are platypuses, along with echidnas (although to a much lesser extent, as we shall see) and a single species of dolphin, the Guiana dolphin, from South and Central America.*

Only recently, in 1986, did researchers establish these electro-receptive abilities of the platypus, and it was nearly a decade later that scientists worked out that they have this gift in order to catch food.[2] Prior to that, we were left to guess at what enabled the platypus to locate food when three of its 'traditional' senses were cut off when underwater – something that was a particular puzzle considering the sheer volume of food they can find: up to half a kilo (one pound) a night. Nevertheless, the bill was always the clear candidate. As Burrell put it in 1927, 'the muzzle of the platypus is possibly the most remarkable organ for sensory perception found in the Mammalia'.[3] His speculations were on the nose, so to speak: 'My opinion is that this animal must have developed some extraordinary means of finding its prey, apart from the sense of touch, and that the sensory apparatus through which this acts is connected in some way with the fleshy nature of the bill.' He called it a

---

* The star-nosed mole – an incredible American species shaped much like the European mole, but with twenty-two long, fingerlike projections forming a star-shaped sensory organ on its nose – had been suggested as another candidate, with the nose-fingers themselves possibly being electro-sensitive, but the results of experiments have so far proved negative. (Bullock, 1999)

'sixth sense', which we now know is the astonishing ability to detect electricity.

As for other senses, platypuses are noteworthy for their lack of whiskers and external ears. Some have cited both of these evolutionary omissions as evidence of the species' primitivity. However, the fold on their heads in which the ear (and eye) sits – the one that is fastened shut when underwater – is highly muscular, with fine controls. The sides of the groove can contract and dilate rapidly, and 'point' at the source of a sound. Much like many other mammal species with mobile ears, it can cock its furrow to act as an external ear. As to the lack of whiskers, it's true to say that this evolutionary sensory development probably appeared after the ancestors of platypuses and echidnas split off from the rest of the mammals, but I think the electro-receptive adaptation more than makes up for it.

In the early 1830s, one of the first Englishmen to keep captive platypuses alive in Australia long enough to observe their behaviours was a naturalist named George Bennett. He wrote that platypuses would allow their bodies to be stroked, but their bills were so sensitive that the animals darted away if they were touched.[4] Another authority on platypus behaviour, David Fleay, reported that they would head back to their burrows in heavy rain 'because the drops of water pattering on [their] highly sensitive bill cause acute discomfort'.[5]

Nonetheless, while underwater, platypuses use their bills to rummage around in the gravel, cobbles and mud at the bottom of lakes and rivers in search of their invertebrate prey. Having caught it, their food is then jammed into their cheek pouches until they return to the surface to eat. There, they crush up the contents of their cheek pouches with the ridged, horny pads inside their bills. Adult platypuses don't have any teeth (three molars appear on each jaw during development but are lost during adolescence), but these horny pads are an excellent adaptation for grinding up the hard exoskeletons of crayfish,

Platypuses doing platypus things – an 1884 print by German artist Gustav Mützel. The skeleton illustration shows the male's venom spurs on its ankles, and the leathery shield at the bill's base has dried into an upright position – typical of historic specimens. In life it lays back across the face.

shrimp and insect prey, as well as for coping with the heavy wearing effects of the sand and grit that comes with foraging on lake and river bottoms. It seems that in their evolution, platypuses have upgraded their feeding equipment from standard teeth: horny pads can be regrown and replaced constantly, unlike most mammals' teeth.

*

Platypuses can live for over twenty years and are found in freshwater habitats across parts of eastern Australia, from Queensland's steamy tropics to Tasmania's icy highlands. The platypuses in the north are much smaller than those in the south. For instance, a big Tasmanian male may exceed 60 centimetres (2 feet) in length and weigh in at 3 kilograms

(7 pounds), whereas a small male from north Queensland may be barely a third that size. This disparity has probably got something to do with it being a lot colder in Tasmania, because larger bodies lose less heat. Female platypuses tend to be about three-quarters the size of males.

Despite the diversity of climates in which they are found, there is only a single species of platypus, and they have a number of behaviours and other adaptations that allow them to occupy this range. Indeed, many of these traits are effective at buffering temperatures at both ends of the scale. Although they are able to sweat, platypuses can die if the ambient temperature exceeds 30 °C (86 °F), but living in a burrow helps with that as conditions are much more stable underground.

Platypus limbs are rather like Swiss Army knives, with gadgets for walking, swimming and digging. To power their swimming stroke, platypuses use their webbed hands (their hind feet are only partially webbed). No doubt this webbing contributed to their overall perceived 'duckiness' and influenced early descriptions, but their webbing is quite different to that of a duck – not to mention the fact that ducks have webbed feet, while platypuses have webbed hands. In ducks, the webs pass between the toes, while in platypuses it passes *under* their fingers. This web also extends well beyond their digits and creates a fan of skin around the hand that is longer than their fingers. Their front claws curve slightly upwards, creating struts to support the webbing during the conversion of the outstretched hand into a paddle.

When swimming downwards, their hands push out, backwards and then up, so that at the end of the stroke their palms are facing upwards by their bodies. This demonstrates just how supple platypuses are. I've tried to replicate this manoeuvre myself, but I can't make it work within the limits of my human flexibility. For the upstroke to reset, the webbing collapses down as it moves through the water, reducing drag.

Platypuses' hands look very different on land, however. There, the webbing is folded back into their balled-up fists, to create a more terrestrial 'foot' for walking, which they do on their knuckles.

Their foreclaws are also put to use as combs when preening, but they mainly do this with their hind feet. Platypuses are extremely flexible and can contort themselves into all manner of shapes when going about regular platypus business. When grooming, they can reach pretty much every inch of their bodies with the claws on their back feet. Keeping their fur in good condition is important for aquatic mammals, and the soft, dense underfur of the platypus is particularly adept at keeping water out. Even after many hours out foraging, a healthy platypus's skin doesn't get wet.

And when digging, they expose their strong claws while the webbing remains tucked away, to act like garden forks cutting into the soil of a riverbank. When you examine a platypus skeleton you can see that their bones have impressive lumps and bumps where large muscles attach, and these give them the strength to create such long burrows.

Burrell believed that the skin of a platypus's hands would need to remain 'cool and moist' to enable this versatility, hinting at constant lubrication – a secretion that foiled one of his less conventional platypus experiments:

> In 1924, while . . . making a moving picture of the natural habits of monotremes, I tried with strong fish glue to fix the staff of an Australian flag in the closed palm of a living platypus, but found that it was impossible.[6]

Extraordinarily considering their body size, a female platypus can dig a burrow that is more than 10 metres (33 feet) long to keep her young safe – with credible claims of burrows exceeding 30 metres (98 feet). The burrows have numerous twists and turns, with side routes ending in dead ends. Burrows that are simply used for sleeping – dug by both males and females – are shorter. Their entrances are typically a little way above the waterline, concealed among vegetation.

Most accounts of platypus burrows mention that they slope upwards from their entrances on the bank towards the surface of the soil, and then run parallel to it, about 30 centimetres (a foot) or less below ground, without emerging into the open. Platypuses avoid digging into the burrows of rakali (native water rats), rabbits, and other bends in their own circuitous tunnels by a similar distance, tunnelling below or around them, but never breaking through.[7] We have no idea how they know those tunnels are there.

Deep inside its nesting burrow, a female platypus will make the nesting chamber comfortable: she collects wet vegetation and drags it to the nest by pinning it between her hind feet and curled tail. This prevents her eggs and nestlings from drying out.

As well as their hands, their bills are also used heavily in burrowing, loosening the soil with solid muzzling of the ground ahead of them, before they insert their hands into the rubble and pull it back over themselves. All the while, the strongly curved hind claws lock the platypus into position in the walls of the tunnel. As platypus tunnels are more or less the same diameter as the animal, turning around would be a challenge mid-excavation. Instead, they can simply retreat back down the tunnel by swivelling their hind feet into reverse: they can point backwards. As Burrell puts it, 'It is rather amusing to witness this act, for, at the outset, the fore-parts are usually obliterated with earth, and the tail, which in contour and elevation somewhat resembles the head, sometimes puts one at a loss to guess whether the creature is really coming or going.' Marvelling at their versatility, he compares their hind feet to the forepaws of a sloth or bear, in that they can form a closed grip as well as performing all this preening and bracing.[8]

*

When you come across any burrow, the shape of the tunnel can give you a good clue about the kind of animal that dug it. Among the most distinctive are scorpion burrows, which look just like an upturned smiling mouth. Their pincers are held outwards and upwards from their bodies, so that's how they dig. Lizards and tortoises hold their legs out in a sprawling posture, so when they dig their limbs push the soil down the sides of their bodies – this means that, in cross section, their burrows are typically arched: flat on the bottom and curved around the top, like a D turned on its side. In contrast, most mammals hold their limbs directly below their bodies, so most of the soil is pushed more centrally under their torsos as they dig. Their burrows are therefore more rounded in cross section. The animals' specific body shape can give more detailed pointers, essentially by picturing the overall front-on silhouette of the creature: rabbit and mouse burrows are more or less circular, foxes' are taller than they are wide and badgers' are wider than they are tall. When it comes to platypuses, their burrows are roughly oval, or sometimes like the sideways 'D', as their legs are held out to the side.

Burrell dug into many burrows and observed countless individuals in the process of forming them in order to provide a detailed account of how they are achieved. Once the platypus has excavated a few inches, it contorts its body to condense the soil down into the tunnel walls and floor, which decreases the chance that they will collapse (they can burrow on their backs and on their sides, and can spin in spirals to tamp the soil down). Very little earth is actually found outside the entrance to platypus tunnels. If soil conditions allow, instead of removing it, it is merely compressed into the walls. There are major energetic advantages to this – 10 metres (33 feet) is a long way to push soil down a tunnel to get it out of the entrance, particularly if you can't turn around. This strategy also helps to avoid detection by predators, both while digging and subsequently: I have regularly located burrowing marmots from the jets of dirt being flung out behind them as they dig (it's also clear that they are having to shove this soil behind them in several stages as they travel further down the tunnel); and the spoil heaps around badger burrows are far more conspicuous than the entrances themselves.

With that said, there aren't too many predators that are likely to be able to attack a platypus by following it up its burrow. Monitor lizards, snakes and rakali are the most likely (although snakes that kill through constriction would have trouble finding sufficient space to wrap their coils around a platypus in the confines of the burrow). Most birds of prey, Tasmanian devils, dingoes and – since Europeans introduced them – feral cats, dogs and red foxes do most of their platypus-hunting above ground. Equally, Murray cod (huge predatory fish), raptors that hunt over water and crocodiles in the north of their range catch their platypus-dinners while they are swimming. Until the recent geological past, Australia also had a suite of large marsupial predators – and who knows which of those would have regularly hunted platypuses?

A characteristic of platypus nursery burrows is that they have a number of little side chambers spaced periodically along their length, budding off immediately next to the main tunnel. Unlike the main chamber, these do not contain any nesting material, so what are they for? Burrell provides a satisfactory answer. They are excavated in order to provide soil to make solid plugs in the tunnel, which the platypus tamps into place as she proceeds along it, effectively sealing her and her young into the burrow at numerous points along its length (they build between two and nine such plugs per burrow, but three is typical).[9] Every time the female leaves her nest, the plugs must be dug through and rebuilt. Coming in from the outside, these plugs are packed so tightly that they are indistinguishable from a solid, unexcavated dead end. This is presumably an effort to stop predators, to maintain the humidity within the nesting chamber or to stop the youngsters wandering off.

These series of plugs must mean that the oxygen level within the nesting chamber dwindles significantly over time, and is only refreshed on the rare occasions that the female leaves. Reading Burrell's work from less than a century ago is an unusual experience – he lists discovery after discovery based on his painstaking observations, and then every so often you are reminded of the platypuses that were sacrificed in the pursuit of his endeavours. On this particular topic he explains that he knew that nestlings could survive extremes of oxygen deprivation because of all his failed attempts at drowning baby platypuses. He sealed one in a bottle of river water for three and a half hours before it eventually died. It's also apparent that he found starving them of oxygen so challenging that on occasion he pickled them in preserving fluid before they were actually dead. I recently found a platypus nestling in a jar of alcohol in our storerooms at the University Museum of Zoology in Cambridge, and was delighted to read on its label that it had been collected by Burrell himself. When I started to wonder whether this had been this individual's fate, it rather took the shine off. It seems that this may have been a common theme in monotreme collecting – I also found a note in a box of echidna microscope slides from 1895, stating that the infant echidna in question had been 'Put alive into strong spirit' by the schoolteacher who had collected it, 'in which it lived for over ¼ hour'.

On a lighter note, in 1835, George Bennett gave us a delightful description of how his young captive platypuses slept curled up in a little ball, much as they presumably do when they are tucked up safe within a burrow: 'This is effected by the fore paws being placed under the beak, with the head and mandibles bent down towards the tail, the hind paws crossed over the mandibles, and the tail turned up, thus completing the rotundity.' Incidentally, he also described the vocalisations of nestling platypuses – they growl when they are dug out of their burrows or disturbed from their sleep.[10]

According to Burrell, 'The adults make a noise which can best be imitated by a tremulous snoring'.[11]

As we all know, curling up in bed is a good way of staying warm (which is even more important for youngsters that aren't fully furred). Platypuses have a lower body temperature than other mammals – a cool 32 °C (90 °F) compared with our 37 °C (98.6 °F) – and their coat gives them excellent thermal insulation. This is essential, as platypuses spend an incredible amount of time in the water. One study of Tasmanian platypuses found that their foraging trips lasted an average of 12.4 hours (with a maximum of a whopping 29.8 hours).[12] Having a cooler body temperature is an important trait to keep down the energetic costs of these long spells in cold water.

Early naturalists thought that platypuses had another adaptation for their aquatic lifestyles. When they witnessed bubbles escaping not from their mouths, but from their backs, they concluded that they were able to breathe through their skin. It would have taken significantly less imagination to determine the true cause: as with all furred aquatic mammals, the bubbles are simply from the water pressure compressing the air trapped in their pelts.[13]

*

Platypus nesting behaviour is, of course, all in anticipation of producing more platypuses. This itself is precipitated by a mating dance that rivals that of any other mammal in its complexity – and it all happens in the water.

The mating season begins when males start to wander beyond their normal foraging patch in search of females, and the scent glands on their chests swell. The females complicate matters by putting considerable effort into avoiding the wandering males (as soon as he turns up, she leaves), until such time as they are ready. As such, females are in control of when platypuses mate. Courtship proper begins when the female initiates a lengthy series of 'swim-bys' with a male, where the pair swim past each other at the surface, checking each other out without making contact.[14]

If they like what they see, over the course of several late-winter or springtime weeks, the relationship gets physical. The female allows the male to bite onto her tail with his beak, and then she leads him in a series of twists and twirls, rolling over, sideways and under through the water. With the female now towing the male, the result is a series of synchronised swimming manoeuvres: soaring, tumbling, freewheeling together, diving, touching and passing. The female will even occasionally swim around and bite onto his tail while he is still latched onto hers, to complete the perfect platypus circle.

Detailed platypus behaviour is hard to study in the wild, so most accounts of this mating dance come from captive animals in large naturalistic pools. One such study, with two females and one male in the pool, observed platypuses turning a love triangle into a love circle, with all three animals biting each other's tails to form a large ring. The two females also formed their own same-sex courtship pair, biting each other's tails and circle-swimming without involving the male.[15]

When the moment comes, captive platypuses can adopt different sexual positions. The male might bite onto the female's back or hold her legs with his front feet, and then curl his tail up under her body. Another mating technique involves the male lying half on his side, half on his back, keeping in position under the female's belly by biting her neck and holding onto her body with his back legs.

As I mentioned, together platypuses and echidnas make up the group of mammals called monotremes. Monotreme means 'one hole', which refers to their cloaca – a one-stop shop through which they do all their defecating, urinating and reproducing. (Cloaca, for what it's worth, is also the Latin word for sewer.)

Until it is required for sex, a platypus's penis is positioned within his cloaca. Upon becoming erect, it pops out, ready for insertion into the female's cloaca. Their reproductive organs differ from most placental mammals in other ways, too. The platypus's penis, which is about 5–7 centimetres (2–3 inches) long when erect, and has a forked head, is only used for delivering semen, and not urine; and their testes are held inside their bodies, rather than in an external scrotum.*

Like humans, female platypuses have ovaries on each side of their bodies. Fascinatingly though, only the left ovary is functional and can produce eggs. This mirrors the situation in birds, where the right ovary shrinks soon after it develops. In birds, it is typically suggested that losing the use of one ovary is a weight-saving adaptation, because it's easier to fly if the animal is lighter. Platypuses can do many impressive things, but flying is not one of them. Not only does this raise the question of why platypuses use only one of their ovaries, but it also potentially casts doubt on whether the reason that birds do the same thing is truly to reduce weight.

After they've mated, the male platypus wanders off and looks for other partners, never to look back, while the female prepares her burrow. In the uterus, platypus eggs develop quite differently to bird or reptile eggs: beginning when the egg is just 4 millimetres (1/8 inch) across – soon after fertilisation – the shell starts being laid down on the egg's surface. This means that nutrients must be absorbed from the wall of the uterus into the egg *through the shell*: the shell actually grows with the egg. Aside from echidnas, no other vertebrates do this.[16] In reptiles and birds, the shell is added after it reaches full size.

If all goes well, three to four weeks later she will lay between one and three – but usually two – eggs. They have flexible shells, containing a lot of the protein keratin, rather than hard calcium compounds, and so appear more like the eggs of snakes, lizards, crocodiles and turtles than those of birds.

Platypus eggs are about the same size and shape as an average

---

*I say 'most placental mammals', as some do keep their balls inside their bodies (a feature known as testicondy), including whales, dolphins, beavers, sloths, anteaters, armadillos and elephants.

marble, but a little less spherical – 17 millimetres (⅔ inch) long and slightly less in diameter. The shells of egg twins or triplets stick together into little clusters immediately after being laid, perhaps helping the mother to keep hold of all of them at once. During the ten days it takes for a platypus egg to hatch, the female will incubate it by holding it against her body with her tucked-up tail. Although platypuses would normally feed for around twelve hours a day – consuming an impressive 30 per cent or more of their bodyweight – during incubation and early lactation, they hardly feed at all.[17] After all that digging, several nights of collecting nesting material and then the energetic strains of producing eggs and then milk, the female must rely on her fat reserves – much of which are stored in her tail. Inside the egg, the platypus embryo grows to about 1.5 centimetres (½ inch) long before hatching. To escape the shell, the embryo has three tools at its disposal: a small, horny egg tooth, pointing inwards from the tip of its mouth; a cone-shaped lump on top of its upturned snout called a 'caruncle', which gives its face the appearance of sporting a massive comedy cartoon human nose; and miniature claws on its hands. All of these are used to help it tear its way out of its leathery shell. Finally, we have the joyous wonder that is a baby platypus.

*

There is no consensus over what a baby platypus should be called. Most writers hedge their bets by going with a generic word like 'infant' or the delightfully Jim Hensonesque terms 'hatchling' or 'nestling'. Baby echidnas – their closest relatives – are called 'puggles'. Because of its obvious charm, people have tried to apply the word puggle to baby platypuses, but there is no real convention for this. So I would like to suggest 'platypup'.

After all, there is no system for agreeing an animal's common name, and certainly not one for the name of an animal's baby. No committee, organisation or government was involved in

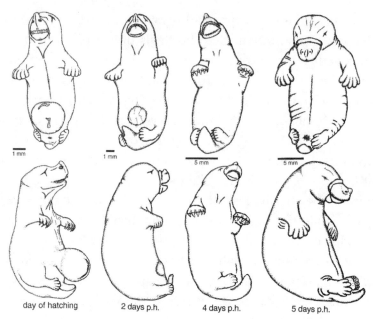

The first five days after a platypup hatches. On day one, the sharp downward-pointing egg tooth can be seen. The caruncle on the tip of its miniature beak is retained for a few months.[18]

the sanctioning of the word puggle for baby echidnas, yet it is now well established (the International Commission on Zoological Nomenclature has the responsibility of agreeing which scientific names are permitted, but not more vernacular epithets). If enough of us decide that platypup is the word for a baby platypus, then so be it. Perhaps if we all drop the word platypup into conversation as often as we can, we will effect a change in the English language. There is a void to be filled, and that void has the shape of a baby platypus.*

*

---

* While we're on this topic, some years ago I began a similar movement to use the word 'wombatlet' for baby wombats. Generally, the word 'joey' is used for baby marsupials of all species, but I find that rather lacking. Wombatlet is now gaining traction, and simply by putting them in this book, it creates a citable reference for the establishment of these words. Please go out into the world and help platy-pups and wombatlets become part of our zoological language.

Held safe against their mother's belly, the platypups will grow, nourished by milk. Unlike adults, nestlings have downward-curved claws on their forepaws to allow them to grip the mother's fur (and they are yet to grow their webbing). These become straighter by the time the youngsters are ready to leave the nest and forage on their own. As I have mentioned, platypuses and echidnas don't have nipples, so platypups drink their milk in a different way to most other mammals. Kittens, puppies and human babies, for example, suckle milk by forming a seal at the front and back of their mouths in order to create the pressure differential needed to draw the milk out of the nipple (to evidence the existence of the seal at the back of our mouths, try to breathe at the same time as sucking). Without nipples, no seal is formed – the oozing milk is simply sucked up from the mother's fur. It's hard to imagine how the beaks of either a platypus or an echidna could have evolved if they needed to form an airtight seal with their lips around a nipple. (That said, whales can't form a seal with their lips either, so jets of milk are squirted out of the nipple into the calf's mouth under pressure; and baby hippos have a similar problem – they form a seal by folding their tongues into a tube around the teat, without using their lips.)

But if there is no nipple to home in on, how does a puggle or a platypup know where to find milk? The mammary glands are found in 'milk patches' on female echidnas' and platypuses' skin, where the milk seeps out. I've seen footage of baby echidnas kneading their mother's milk patch as they feed, presumably to encourage the release of milk. Platypups are thought to do the same thing. Indeed, kneading of the breasts by infants is also seen in humans and cats. What's more, the platypus's unique caruncle – the knob on their snouts that helps them rip their way out of their eggs – doesn't disappear for a few months after hatching (unlike the egg tooth, which is gone after a couple of days). Researchers suggest that platypups use it for rubbing the mother's milk patch in order to stimulate the production of milk.[19]

When they don't have young to feed, a platypus's mammary glands are only around a centimetre (⅓ inch) long, but when needed they grow to cover most of her underside, and even spread up and around onto her back (all under the skin). Little ducts lead down to the small milk patches on her belly.[20] The speed at which young platypuses grow must require an awful lot of milk.

In their burrows, platypups may be energetic – one of George Bennett's correspondents described them as 'most playful little things, knocking each other about like kittens, and rolling on the ground in the exuberance of their mirth'.[21]

When the babies are large, and demand for milk is highest, the female can consume an astonishing 90–100 per cent of her body weight in food each day.[22] Think about that. The average British woman weighs 70 kilograms (11 stone, or 154 pounds). Breastfeeding humans have an increased appetite, but imagine eating 70 kilograms of food in a day.

Platypuses lactate for three to four months. As far as is known, only then will the platypups leave the burrow for the first time and begin life as a fully fledged platypus. However, the fact that from a very early stage, young platypuses' hands are capable of gripping – and Burrell's horrific stories of how hard it is to drown a platypup – has led some to speculate that perhaps the babies cling onto the mother's fur when she briefly leaves the burrow to feed.[23] That sounds like a very risky strategy to me, particularly as the youngsters have no fur during those early stages, and so would be very susceptible to the cold of the water. In any case, all that milk in the burrow means that young platypuses emerge at an impressive 80 per cent of their adult length (that's the equivalent of a human baby reaching 1.2 metres (4 feet) tall by the time they stop breast-feeding) and two-thirds of their adult weight.[24]

*

The list of venomous mammals is almost as short as the list of electro-receptive ones. It comprises slow lorises, platypuses, a few species of shrew and the Caribbean solenodons, which are related to shrews. Shrews create stores of food to ensure they always have enough fuel to maintain their super-fast metabolisms, and it is assumed they use their venom to keep their prey alive but in a comatose state to keep it fresh. Another less gruesome possibility is that being venomous allows the minute shrews to attack and kill prey that is larger than they are.

Meanwhile, the slow lorises – small Asian primates with flat faces, large eyes and long, flexible bodies for twisting between tree branches – normally use their venom for fighting amongst themselves. However, a friend of mine, Australian ecologist George Madani, is one of the few Westerners to have been on the receiving end of a venomous loris bite. From his experience it's clear that loris venom also has a defensive function. In an experience somewhat reminiscent of one recorded by Karl Patterson Schmidt, the celebrated curator of zoology at the Field Museum in Chicago, George's first thought on being bitten was to write a scientific account of the symptoms he was suffering. In Schmidt's case, he was bitten by a boomslang snake in 1957 and recorded a blow-by-blow description of the biological effects it had on his body. As his condition deteriorated, it was suggested that Schmidt should seek urgent medical attention, but he refused to go to hospital, fearing that it would invalidate his account of the symptoms. Schmidt insisted the snake was too small to produce enough venom to kill him. He was wrong.

Happily, my friend George did not die. Not only did he suffer from the toxicological effects of the slow loris bite, however, but the venom also induced a severe anaphylactic shock. His face – particularly his lips and tongue – swelled to a point where it seemed as though he might be unable to breathe. He did make a full recovery, but the moral here seems to be: don't mess with lorises.

A venomous animal is typically defined as one that produces a toxic chemical that it injects into another animal, thus disrupting the biological processes of that animal. Conversely, animals that produce toxins but do not inject them are termed poisonous (they become dangerous if you ingest the toxins or they are absorbed through your skin), like some frogs, puffer-fish and many beetles and butterflies. By this definition, the only other mammals that are sometimes added to the 'venomous' list are vampire bats, as their saliva affects the ability of the animals they bite to form blood clots around the wound.

In 2019, I curated an exhibition of artworks by the legendary zoologist Jonathan Kingdon, who was born in Tanzania in 1935 and has spent a lifetime observing the behaviour and anatomy of wild animals across Africa. It was a true pleasure to discuss his life with him as we worked towards the exhibition. Kingdon has made countless discoveries about how and why animals look and behave the way they do, and then translated these discoveries into stunning works of art. At one point, learning that the dogs belonging to the local people he was working with would die if they caught African crested rats, Kingdon investigated the possibility that these rats too were venomous, and that they were biting the dogs in defence. What he discovered, however, was that the rats seek out and chew up the toxic bark of poison arrow trees (so-called as the bark has traditionally been used to tip hunters' arrows in order to kill elephants), and then lather themselves with their now-toxic saliva. Crested rats have grooves in their hairs which make them absorbent, so their fur soaks up the toxins to provide a highly potent defence against any predator that bites them. The rats also have a striking orange, black and white striped colouration, which acts as a warning signal, much like orange bands on a wasp. These rats are therefore poisonous, but not venomous. Hedgehogs are known to do something similar: they can smear the toxins that toads produce in their skin over their spines, adding an extra weapon against anything foolish enough to bite them.

When it comes to platypus venom, it does play a defensive role (which, as we shall see, both humans and dogs have learned the hard way) but that is not its main purpose. That only male platypuses can produce venom demonstrates that defence cannot be its only use, because if it were, females would be venomous too. Its key purpose is that the males use it to fight with other males over territory and mates.

In fact, platypuses are the only animals whose venom production is known to be seasonal, focused as it is around the breeding season. Most venomous vertebrates – like snakes and lizards – produce their venom in their mouths and deliver it with their teeth.* In platypuses, however, the venom glands sit under the skin near the hips, and connect via a channel to 2-centimetre long (1 inch) curved, conical spurs near each of their ankles. These spurs are composed of two small articulating bones covered by a horny sheath. Females are born with the spurs – but not the venom system – and they fall off as they grow during their first year. Male echidnas also have spurs (again, they fall off during the female's development), but they do not produce venom. Instead, the glands attached to their spurs produce a milky white substance that is assumed to be used in attracting mates or to induce the female to ovulate.

Platypus spurs are pointed towards each other. To put them to use, the male erects them using strong muscles, wraps both hindlegs around his victim and drives the spurs in. He injects a few millilitres of clear, toxic liquid through repeated jabbing. Platypus venom is not usually fatal to other platypuses,† but

---

* Incidentally, slow lorises produce some venom in their mouths, but also produce another toxin from glands on their arms, and lick this up before delivering both toxins to their victims through the attacking bite.

† Although it can be. David Fleay described finding a young male that shared its zoo enclosure with an 'old man' platypus dead one morning 'and as full of holes as a colander' (Fleay, 1980).

it gives the stabber enough time to mate with a female while the stabbee is temporarily immobilised.

With natural platypus behaviour being so hard to study, the purpose of their spurs – and whether they carried venom – has been controversial. Much of the evidence for its assumed use in male-on-male fighting is based on finding wounds on male platypuses that look very much like they could have been inflicted with a spur. We do not have a better theory than that.

Chief platypus-wrangler Harry Burrell had something to say on the matter. He, like others before him, agreed with a theory first suggested in 1802 by Everard Home (pronounced 'Hume'), in the species' first full anatomical description: that the males' spur was used to hold the female during mating (sharks, for example, have claspers near their pelvic fins for securing their position during sex). He suggested that the spurs are well positioned to be hooked around the females' legs 'in the season of love'.[25]

As I mentioned, in females, the spurs grow but are shed when young. In their place are remnant marks of the lost appendages, which Burrell described as significant indents. He went on to suggest that the male's spurs slotted into these patches where the female's spurs had once been.

Burrell appears rather confused about the possibility that platypuses are venomous, given his conclusions regarding the spurs' use. He is very clear that there is a gland attached to it, and that there is a small hole a little way down the curve of the spur where the duct releases a secretion to the outside. But what is that secretion? At one point he shares a theory that if the male did insert his spur into the scar where the female's spur once was, the tessellation would be so tight as to hinder his eventual escape. Could it be that the liquid that the spur produces is a muscle relaxant that allows it to be released from the female? Burrell shares a lot of the evidence available in 1927 that strongly suggests platypuses were venomous, without seeming entirely convinced that it was true.

He quotes Sir John Jamison's 1816 letter to the Linnean Society of London, one of the first accounts of the platypus's potential for pain:

I wounded one with small shot; and on my overseer's taking it out of the water, it stuck its spurs into the palm and back of his right hand with such force, and retained them in with such strength, that they could not be withdrawn until it was killed. The hand instantly swelled to a prodigious bulk; and the inflammation having rapidly extended to his shoulder, he was in a few minutes threatened with locked-jaw, and exhibited all the symptoms of a person bitten by a venomous snake. The pain from the first was insupportable, and cold sweats and sickness of the stomach took place so alarmingly, that I found it necessary, besides the external application of oil and vinegar, to administer large quantities of the volatile alkali with opium, which I really think preserved his life. He was obliged to keep his bed for several days, and did not recover the perfect use of his hand for nine weeks.[26]

Others who have similarly suffered from grabbing a male platypus during the breeding season have given comparable accounts. One report described how, when a platypus locked both spurs in, the grip was so strong that two men were required to pull it off its victim.[27]

The venom is not fatal to humans, although dogs have died from it. But its effects are unusual, particularly as it causes pain throughout the whole body as well as around the wound area, along with massive swelling over a large area followed by muscle wastage. The effects can last for months or even years, with some victims suffering discomfort and stiffness in the area they were stung long after the incident. The pain is said to be excruciating and cannot be eased with general first aid or painkillers. In one account, the victim – a decorated war veteran called Keith Payne – said that the pain was far worse than his experience of shrapnel wounds. He had seen a platypus sitting near him on a log while he was fishing, and because it didn't move when he approached, he assumed it was sick or injured and picked it up. It stabbed him in the hand and middle finger. The ensuing pain lasted for over a month, and fifteen years later he reported that he was still experiencing discomfort and stiffness, and his use of the affected hand was restricted. There is currently no antivenom. However, it is thankfully rare for people to be attacked by a platypus.

In trying to empirically determine how the venom worked, Burrell experimented with the reckless abandon of early twentieth-century science. He writes of a frankly horrendous trial involving injecting a tiny amount of venom (a solution of 0.06 grams of dried powder) into a rabbit's jugular:

> Within three seconds from the commencement of the injection the blood-pressure fell by 40 mm. of mercury, the heart-beats becoming less frequent. At the same time the respiration became hurried and exaggerated, and speedily terminated in a series of expiratory convulsions, in the course of which the blood-pressure rose again, but speedily fell. In a minute and a half the animal was dead.[28]

We now know that the make-up of platypus venom is extremely complex, comprising around twenty different components, each of which has different, unpleasant impacts on the victim. Some induce pain, while others cause low blood pressure and oedema (swelling caused by a massive build-up of fluid) by opening up unique channels in cell membranes. Another is similar to a component of snake venom that makes cells more susceptible to the other toxic components. Yet another has a highly unusual chemical structure that is extremely resistant to modification, which means it is very stable and could explain the long-lasting effects on victims.[29]

Toxins from other animals have been used in the development of a number of drugs, and the unusual nature of platypus venom means that it could potentially help unlock discoveries about the nature of pain and what is actually happening inside pain sufferers at a physiological level. It also opens up possibilities for research into new painkillers that target different biological pathways to existing drugs.[30]

Yet although platypus venom is unique, there are structural similarities in some of its components and their effects to the toxins of other animals, such as spiders, frogs, snakes, shrews, fishes and sea anemones. The venom components seem to have evolved separately in each group, but evolution has repeatedly built proteins containing the same or similar specific structures as templates for the venom molecules. It is an incredible example of convergent evolution, where similar biological solutions are reached independently in different animal groups to solve similar problems.

Intriguingly, discoveries in the fossil record have raised the possibility that the very first mammals were venomous. Palaeontologists have found tiny ancient mammal fossils from the Early Cretaceous Period – around 125 million years ago, meaning that they are older than the oldest known monotremes – with minuscule bones near their ankles that correspond to the bones that support platypus venom spurs.

That these bones have not been found in mammals' closest fossil relatives suggests it is a uniquely mammalian trait. It also means that rather than having been acquired during monotreme evolution, these spurs are features that platypuses and echidnas have retained from their early mammal ancestors, while marsupials and placentals have lost them.*[31]

*

---

* This is important, as the presence of spurs has often been used as a unique defining feature of monotremes; however, if it's a feature that is general to mammals (but has been lost in some groups) it cannot be used to define the monotreme group.

The vast majority of my own encounters with platypuses have taken place in Tasmania, and as it happens, that is where I was when I began writing this book, during fieldwork in August 2019. Since 2010, I have been assisting Rodrigo Hamede Ross and his colleagues at the University of Tasmania with their research into a horrific contagious cancer that is decimating the Tasmanian devil, the largest surviving marsupial carnivore. It is an incredibly sad story. In 1996, a devil was captured with facial tumours in the far northeast of this heart-shaped island. Since then, the disease it carried – devil facial tumour disease (DFTD) – has rapidly spread across nearly the entire state. It is passed from devil to devil by biting, something they habitually do to each other during many social interactions.

Devils are amazing creatures – stocky, meaty, squat and strong, like a cross between a badger and a pit bull. Their fur is black, normally with irregular patches of white on their throats and across their rumps, with large pink ears and long, fat, tapering tails. Their broad muzzles are sparsely furred, and their faces have several areas of extraordinarily long whiskers – sometimes approaching their elbows. These likely have several functions, including guiding them through dense bushes and between log and boulder piles where they den, but also letting them know how close they are to other devils when feeding around the same carcass, and their heads are buried in their meals. Devils are primarily scavengers, with an excellent sense of smell. There are plenty of dead animals to eat in Tasmania, and devils are very well equipped to do so. They have broad, bumpy molar teeth and massive jaw muscles for crushing bone. Although most of their relatives are solitary, if a carcass is large more than ten devils can crowd around it to feed, which leads to vocal squabbles and fights. It is these growls, screeches, banshee wails and gurgled howls that give devils their English name. Being woken by these demonic sounds in the middle of the forest at night is spine-chilling (even when you know what they are). I've seen devils chew their way into a wallaby's anus and climb most of the way inside in an attempt to reach the choicest parts of the carcass. Hair, claws and bone shards are all swallowed down (and can emerge back out the other end), and after the devils have done their work the entire carcass is typically gone.

As they age, particularly the larger males, their faces become heavily battle-scarred from all the brawling. Devils must have incredible immune systems in order to stave off infection from these wounds. When we trap them, we sometimes find individuals with 15-centimetre (6-inch) gashes across their throats and missing noses and ears. Sadly, however, devil facial tumour disease has found a weakness, though it is not currently known how it gets past the devils' immune responses.

DFTD is not caused by a virus or mutagen; it is contagious because the cancer cells themselves have become the pathogen. When a diseased devil bites another devil, cancer cells can rub off onto that second devil, and those cells can grow into new tumours. These tumours can grow extremely large and ulcerated – they are upsetting to come across. They are mostly found around the mouth, resulting in teeth falling out or even much of the meat of the face being eaten away. Strangely, although devils do bite each other all over, it is very rare to find tumours growing anywhere other than the face. Within a year of contracting the disease, most devils will die of organ failure, starvation or secondary infection from weakened immune systems.

On my first field trip, in 2010, we caught, studied and safely re-released devils 243 times in ten days. We would set out before dawn and often there were so many devils to process and release that we wouldn't return until after dark. These long days were some of the most wonderful experiences a zoologist could hope for, all set in the truly stunning environment of the Tasmanian highlands. Since then, we have set the same traps in the same places, but the number of devils decreases every time I go. In 2019, it had plummeted from 243 to 9.

One consequence is that it no longer takes us the whole day to check the traps and process the animals, because nearly all of them are empty. This leaves me much of the afternoon to follow wombats across the mountainsides, and the evenings and nights to search out the most fruitful nearby spots for platypuses.

The most likely place – where a sighting is basically guaranteed – is a little pond by the parking lot of a 'wilderness hotel' near the field site. But this almost feels like cheating: it is definitely wild, but the proximity of the hotel and its parking lot makes it feel like a zoo. Instead, I find platypus-watching by rivers more rewarding, as it's harder. Any local platypus isn't confined in one place and you don't have the benefit of looking for disturbances in the stillness of a lake's surface. However, the payoff for sitting, freezing on a riverbank for hours is typically only a few seconds of viewing time when a platypus swims past in the current at speed.

As luck would have it, one of the other members of the team on the 2019 devil-monitoring trip was a platypus researcher, Jana Stewart, who, like me, was taking some time out to help with the devils. We spent a lot of time perched on a mossy log, staring at the ripples in a nearby rainforest creek, whispering about the latest platypus research while we waited for one to pop up.

Platypus research is diverse. For example, one tragic discovery, she told me, is that platypuses in the state of Victoria have been found to have high levels of antidepressants in their blood, because it is being passed into watercourses from human urine. More positively, we talked about how platypus milk has the potential to help in the fight against hospital superbugs. In other mammals, the nipples act as sterile delivery devices for the milk: it goes straight from the mother into the infant's mouth without ever being contaminated by the outside world. But as platypuses don't have nipples, with the milk sort of sweating out onto the mother's skin where the fur wicks it up and the babies suck that, they need another means of sterilising it. The milk, it turns out, has considerable antibacterial properties, and this could now be utilised to solve the global challenge of antibiotic resistance in humans.[32]

This, and the possibility of their venom being useful for the development of painkillers, are not the only ways in which platypuses have been considered as potentially important for human medicine. A beautiful specimen on display at the University Museum of Zoology in Cambridge was brought into the UK in the 1970s (freshly dead – given to the then president of the Royal Society as a memento of his visit to Australia) for the specific purpose of researching the structures of its proteins and comparing them to those in humans. The idea was that looking at the sequence of amino acids that make up the complex molecule that carries oxygen within muscles (called myoglobin) in other mammals – particularly one so distantly related as the platypus – might help unravel human diseases caused by changes in the way that protein is built.[33] After taking muscle samples, the museum's then curator of vertebrates, Adrian Friday – who was one of the researchers studying the proteins (and, incidentally, the man who first properly introduced me to platypuses when I was his student some thirty years later) – freeze-dried the specimen for display in the museum.* This is a preservation technique that removes all the moisture from a specimen (so that it doesn't rot) through freezing and desiccation in a vacuum chamber. The result appears like taxidermy on the outside, but unlike taxidermy, the animal's insides remain in place, just desiccated.

Back to Tasmania. One evening, Jana and I had been in the wilderness hotel's bar to make use of the Wi-Fi (having a bar *and* Wi-Fi a couple of miles' walk from your cabin is a rare luxury on ecological fieldwork in Australia), and

---

* In fact, it was this individual specimen that kick-started my passion for platypuses.

we set out as dusk fell to head to our log on the nearby river. However, as we passed that pond by the parking lot, a platypus surfaced.

As a wombat wombled in the background (a truly wondrous zoological union) we watched the platypus make a series of foraging dives, with ten-second interludes on the surface. As much as it seemed odd to abandon a platypus there in front of us, the hotel lights, parked cars and tourists made it all seem a little canned, and on top of that the lake water was opaque with peat. By contrast, if a platypus showed up on the creek, the water was clear and shallow enough to watch it hunt. So, we decided to leave our 'bird in the hand' behind and search out a wilder-seeming version on the river while there was still enough light to see.

The mass of snow from the previous days had nearly disappeared and the river was high and fast with meltwater. It was hard to imagine platypuses swimming well in these conditions, and we chatted about a study conducted nearby on the physiology of platypus locomotion.

Admittedly, platypuses seem to be awkward movers. Most vertebrates that swim push themselves forwards through the water using their hind limbs or tail, whereas platypuses are unusual among swimming mammals in that they row through the water using their front limbs. Polar bears are one of the few other mammals to do this,* which isn't a comparison that comes to mind easily.

Platypuses hold their limbs out sideways from their bodies while paddling or walking, and move their bones in a unique way. This is because platypus life is essentially a series of compromises. For one thing, as well as walking and swimming, their limbs are also adapted for digging, meaning that unlike most swimmers, their legs are heavier at their hand and foot ends than they are at their shoulders and hips.

---

* As do fur seals and sea lions.

When animals live across these multiple lifestyles, they tend to be less efficient at each thing that they do than species that specialise in just one activity. For example, semi-aquatic animals are assumed to expend more energy while walking or swimming than either specialist terrestrial or aquatic animals. This is true for semi-aquatic placental mammals like the rakali and mink. And by sticking a platypus on a treadmill in a respiration tank, researchers at the University of Tasmania have found that it is also true for platypuses.[34]

However, they also found that the platypus's energetic costs for walking, swimming and resting were *half* that of placentals of a similar size and lifestyle. Once again, this shows that platypuses are brilliantly adapted to their way of life.

With all that said, Jana and I didn't see a platypus on the river that night. Maybe the water was flowing too fast for them so they stayed in their burrows, or maybe one zipped past at speed and we just didn't spot it. The gamble we made by switching locations didn't pay off, but it did give us a chance to really think about how these animals live their lives in their environments.

*

If I sound excited about platypuses, it's because I am. Despite the idea that science is supposed to be an empirical, systematic study of the world, conducted through observation and experiment, the people doing those observations and experiments are not, on the whole, devoid of emotion. We scientists do what we do because we are enthusiastically interested in our subjects. This means that, while we work extremely hard to stop prejudices and subjectivity from clouding our interpretations, science is rarely truly impartial or wholly objective. No one studies animals without passion. Wouldn't it be strange if they did?

However, putting that emotion aside, scientifically speaking,

I still think platypuses are the most amazing animals on Earth: they are an evolutionary biologist's dream, as they physically embody precisely how evolution works. They have features that they share with their ancient reptile-like ancestors that lived over 300 million years ago. For instance, they lay eggs and their shoulder bones are arranged more like they are in reptiles. But on top of this, other features have appeared in platypuses' evolutionary history that are almost unheard of in other modern mammals, like the versatility of their front feet and their detection of electricity. Evolution can only work with what it's got, and then add and subtract from there. In the platypus's case, novel features have been added without losing some of the traits they inherited from their ancient predecessors.

To better understand where platypuses – and all other mammals – came from, picture a typical packet of plastic toy dinosaurs. It probably contains an animal with four bent legs, a huge sail on its back and a set of impressive teeth. Contrary to popular belief, this creature isn't a dinosaur at all – it's more closely related to humans than it is to dinosaurs. *Dimetrodon*, as it's known, lived around 295 million years ago – long before the dinosaurs – and belonged to a group of animals called therapsids, from which mammals evolved.

Dimetrodon and other ancient therapsids broadly looked like reptiles, and so the first mammals – appearing from within the therapsids – would have also looked like reptiles. Certainly, they would have shared a lot of features that we consider reptilian, like laying eggs. In the time since the first mammals evolved, nearly all living mammals have been the result of evolutionary pathways that involved changing those 'reptilian' features, but the path that led to modern monotremes did not. Instead, they have retained some of these ancient characteristics.

That evolutionary path was long and winding, and there are frustratingly few fossils to help us understand the deep history of monotremes. The first mammals evolved around 210 million years ago, which is only slightly after the first dinosaurs appeared. The lineage that led to monotremes then branched off from other mammals at least 180 million years ago, although the oldest known fossils that are believed to be true monotremes are around 125–110 million years old. A small handful of animals, including species called *Teinolophos* and *Steropodon*, have been described from Australia from this time. The only evidence of their existence is a few teeth and some tiny fragments of jaw. This doesn't give palaeontologists much to go on, but, based on similarities with the minute teeth which develop in young modern platypuses (but which disappear before adulthood), both are thought to have looked to some extent like platypuses. *Teinolophos* was about the size of a mouse (but probably lacked the platypus bill), whereas *Steropodon*, at about the size of a cat, was one of the largest mammals alive during the Cretaceous Period. The latest research suggests that monotremes' ability to detect electricity appeared very early in the group's history, when they were evolving near the South Pole. This adaptation would have enabled them to hunt in the darkness of the polar winter.[35]

Intriguingly, the next oldest known monotreme fossils are a few fossilised teeth and a bit of thigh bone that were found in Argentina. They represent two species – one was described in 2023 from just before the non-bird dinosaurs went extinct 66 million years ago, named *Patagorhynchus*. The other, *Monotrematum sudamericanum*, is just a little younger, from after the asteroid hit. These animals are believed to have been species of platypus.[36]

And so, ancient platypuses once lived across the continents of Australia and South America – and presumably Antarctica – when these giant landmasses were connected to each other. That is how platypus ancestors got from Australia to South America. Antarctica was not covered in snow and ice at this point, but had a similar climate to parts of Tasmania and

South America today. Indeed, this is the means by which marsupials reached Australia: they evolved in the northern hemisphere, crossed from North America to South America during one of the previous occasions when these two continents were connected, then went across Antarctica and up into Australia. The last land bridge to fall was when Australia disconnected from Antarctica around 35 million years ago.

A handful of fossils from the last 26 million years have been discovered in Queensland and South Australia that represent species in the genus *Obdurodon* (a genus is a group of closely related species). Like the earlier fossils, this genus was also first described from fossilised teeth (*Obdurodon* means 'enduring teeth'), and the subsequently discovered 15-million-year-old skull of a species named *Obdurodon dicksoni* bears a striking resemblance to modern platypuses, except for the large crushing teeth it has at the back of its jaws. When palaeontologists discovered a single, large platypus tooth in 2012, the press went to town on stories of a 'giant extinct platypus'. Based on just that one tooth, which was between 5 and 15 million years old, the researchers scaled the animal up to around a metre (3 feet) long, roughly twice the size of a modern platypus. They accompanied the announcement of the discovery with an artist's reconstruction of the beast crushing a freshwater turtle in its bill. It was named *Obdurodon tharalkooschild*, in honour of Tharalkoo, a duck who gave birth to the platypus in an Aboriginal origin story.[37]

The oldest fossils that are thought to belong to our modern-day platypus are 3.8 million years old.[38]

*

So, today the platypus is the only one of its kind, and although the final stanza is not technically true – since all animals are related – I think the spirit of Banjo Paterson's poem *Old Man Platypus* endures:

Far from the trouble and toil of town,
Where the reed beds sweep and shiver,
Look at a fragment of velvet brown –
Old Man Platypus drifting down,
Drifting along the river.

And he plays and dives in the river bends
In a style that is most elusive;
With few relations and fewer friends,
For Old Man Platypus descends
From a family most exclusive.

He shares his burrow beneath the bank
With his wife and his son and daughter
At the roots of the reeds and the grasses rank;
And the bubbles show where our hero sank
To its entrance under water.*

Safe in their burrow below the falls
They live in a world of wonder,
Where no one visits and no one calls,
They sleep like little brown billiard balls
With their beaks tucked neatly under.

And he talks in a deep unfriendly growl
As he goes on his journey lonely;
For he's no relation to fish nor fowl,
Nor to bird nor beast, nor to horned owl;
In fact, he's the one and only![39]

*

---

* The pedantic scientist in me can't help but point out that male platypuses don't share their burrows, and burrow entrances are above water.

'How come platypuses are considered mammals,' people often ask, 'given that they lay eggs?' It is a common misconception that a defining feature of mammals is that they give birth to live young (this is one of the reasons that pre-Darwinian taxonomists had so much trouble working out where platypuses fitted on the tree of life).

As the first mammals laid eggs, giving birth to live young cannot be a defining feature of mammals. And this means that we need not expect all mammals to give birth to live young today.

Instead, among the things that *do* unite all mammals is the fact that the females feed their young by producing milk from mammary glands (which is where the word mammal comes from). They also have fur and a single bone in each side of their lower jaws (reptiles typically have seven – parts of the reptilian jaw shifted backwards to form the ear bones in mammals). Mammals have a unique arrangement of bones in their ankles, and they can raise and maintain their body temperatures above the ambient temperature.* If an animal has these features or has evolved from an animal that had them (for example, adult whales have lost fur and ankles during their evolution, but their ancestors had them), then they are defined as mammals. Platypuses and echidnas have all these features.

As I've mentioned, mammals are split into three major evolutionary groups: monotremes, marsupials and placentals, each of which is characterised by how they reproduce (eggs; relatively tiny newborns; relatively larger newborns). Each of these groups also has defining features that we can spot on their skeletons. This makes life easier for palaeontologists who

---

* This is officially known as being homeothermic – scientists tend to disapprove of the popular term 'warm blooded', because animals that aren't homeothermic can also have 'warm' blood. Lizards and frogs, for example, are not 'cold blooded': scientists call them poikilotherms, meaning that they need to get their body heat from the external environment, such as being heated by the sun.

might struggle to detect an animal's reproductive strategy from fossils alone. After all, if the only way to tell the difference between them was how they produced their young, we wouldn't know how to classify any male mammal fossils. For example, monotremes have additional bones in their shoulders, whereas marsupials have a little shelf on the inside of their lower jaws and placentals lack the two struts of bone, called epipubic bones, that the other groups have sticking forwards from their pelvis.

*

There are currently only five recognised living species of mono-treme (four, or possibly three, kinds of echidna and the single platypus), and fewer than 380 species of marsupial.[40] The rest of the approximately 6,500 mammal species are placentals. This can lead to marsupials, and monotremes especially, being considered the odd ones out, while placentals might seem to be 'normal' mammals. But just because they are in the minority doesn't make them any less mammalian.

I expect that most people think of platypuses affectionately. Nevertheless, it's disappointing to see how words like 'weird', 'bizarre' or 'curious' are used pretty much universally in the articles and documentaries that feature them. I suspect that the weirdness-factor is entwined with their popularity, and is exploited for clickbait, TV viewing figures or whatever the platypus might be being used to sell (they have appeared in adverts for Pringles, banks and hybrid electric cars, among other things). Platypuses are popular because they are inter-esting, but that doesn't mean we ought to introduce these lazy value judgments. Reminding me of these prejudices on a daily basis, I have a coffee mug with a drawing of a platypus on it. In gold letters it says, 'Keep it WEIRD'.

Novelty crockery is one thing, but as I was writing this, my monthly *BBC Wildlife* magazine arrived through the letterbox.

It's the UK's leading popular natural history publication. The platypus was the cover star that month. 'Stranger things', a headline says, 'Up close with nature's weirdest mammal'.[41] Groan. The stories *BBC Wildlife* was telling about platypuses were accurate, but if it were me writing, I would want the readers to come away with a sense of how amazing these animals are, not that they are weirdos. Similarly, the BBC's 2019 landmark documentary series *Seven Worlds, One Planet* included this line in the description of the episode exploring Australia: 'Isolated for millions of years, the weird and wonderful animals marooned here are like nowhere else on Earth.' However awe-inspiring and conservation-focused the production was, the framing of Australian wildlife as weird is unnecessary and unhelpful. In the first half of 2020 alone, *The New York Times* ran two stories on platypuses. One had the headline 'Can the World's Strangest Mammal Survive?' and the other included a caption describing them as 'Dr Frankenstein's first attempt'.[42]

These descriptions are intended to be celebratory, but they make platypuses and other Australian mammals seem intrinsically alien. They dismiss them as evolutionary curiosities. The words have serious consequences: they imply that they are just funny little beasties, amusing and interesting, but not important in the scheme of things. If you think about it, pretty much every animal species could be considered bizarre. Elephants, rhinos, bears, deer, whales and so many other animals are unquestionably odd, but they are far more likely to be described as 'majestic' than 'weird', and they have extraordinary societal value placed on them.

As I have said, platypuses are also commonly described as primitive, and that simply doesn't make scientific sense. Firstly, no living, complex species can be considered primitive, as they are all equally evolved. In evolutionary biology we do have the notion of 'primitive features' (also known as ancestral features), which are specific attributes that a species has inherited from its ancestors without much modification. It's a relative term used to distinguish between features that have only recently evolved in a specific group from ones with a deeper history. In this way, laying eggs is said to be a 'primitive' feature, which platypuses share with reptiles. From there, people make the leap to describe the *species* as primitive. But then birds lay eggs, too, a feature they also inherited from their reptilian ancestors, dinosaurs. So why does egg-laying make platypuses primitive, whereas the term is never applied to birds?

Humans share lots of 'primitive' features with reptiles, too. For one thing, we have legs. This is a feature we – like all mammals – inherited from the common ancestor we share with reptiles (in fact we can trace it back to our evolutionary history among the fishes). That we protect our developing embryos in a liquid-filled sac surrounded by a membrane called an amnion (that's the membrane that bursts when the waters break during labour) could also be seen as primitive in the same sense. The point is that all species have primitive traits, but that does not make an animal primitive.

Simply put, the platypus is most famous for laying eggs, and the reason that this is widely considered as primitive is because it is an inherited feature that we humans do not have. Thinking of the platypus as primitive is therefore a manifestation of the idea that humans are the pinnacle of an evolutionary ladder. We are not. Evolution is not directional: the last four and a half billion years of Earth history have not been leading steadily towards the recent appearance of humans. We are just one species on the tree of life, as is the platypus. We humans may be relatively good at things like karaoke and reading books, but we are utterly lame at finding and catching worms buried in the sediment at the bottom of a lake with our eyes closed. We are the best at being humans, and platypuses are the best at being platypuses.

2

Diplomatic Platypuses

Since the medieval days of the royal menagerie at the Tower of London, and even before then, animals have been used as political gifts from one country to another, as a means of fostering goodwill and cementing international relationships. Perhaps 'panda diplomacy' is the best known of such endeavours – the gift or loan of pandas by the Chinese government (particularly between the 1950s and 1980s) to countries with which it wanted to reinforce closer relations. The same was tried with Australian egg-laying emissaries to the UK and America during the early- to mid-twentieth century – an act that has been described as 'platypus diplomacy'.[1] Kangaroos – unquestionably an Australian icon – had been exported to London since the earliest days of the colony, but not as a result of deliberate government action. The platypus, however, was so imbued with the iconic distinction of being uniquely Australian, was the subject of so much scientific interest, was rare in captivity and hard to keep alive (and yes, exhibited a happily exploitable popular peculiarity) that its gifting carried sufficient cachet to represent a political statement. The acceptance of such a gift had meaning, too, as it was a significant commitment to keep them fed.

Australia had allowed several iconic species 'out' before this time, including thylacines (which are now extinct), Tasmanian devils, koalas, echidnas and others, but agreeing to export platypuses to another country was a true rarity.

In 1943, in the midst of World War II, Winston Churchill asked the Australian Prime Minister John Curtin for the remarkable gift of six live platypuses. He thought that the appearance of these marvellous creatures in the UK would be a national morale-booster and improve soured relations between Australia and Britain. On the basis of the difficulty of such an undertaking, he was persuaded to settle for one platypus (which was also to be called Winston), and David Fleay of Healesville Sanctuary in Victoria was charged with supplying it and organising the mission, all in secret. Fleay had a way with words: 'Imagine any man carrying the responsibilities Churchill did, with humanity on the rack in Europe and Asia, finding time to even think about, let alone want, half-a-dozen duckbilled platypuses'.[2]

It was felt that when Japan had entered the war, Churchill had prioritised the battles in North Africa and the Middle East by being slow to allow Australian forces to return to defend themselves in the Pacific. Perhaps by agreeing to Churchill's request, a platypus could heal such wounds and bring Australia's national considerations higher up in his priorities. However, despite having been built a state-of-the-art platypusary (real word) and fed 700 worms a day on board a well-armed ship, the duck-billed Winston died just four days before arriving at the docks in Liverpool, northwest England. A submarine attacked his ship. It didn't sink, but the shockwaves from the defensive depth charges were presumably too much for an animal with sensory systems as acute as a platypus's, which had also been weakened by a reduction in his worm rations due to dwindling supplies.[3] Although the 'mission' was kept secret from the British public (the failure would not have looked good), the dead specimen ended up taxidermied on Churchill's desk. Its whereabouts are currently unknown.

Although Winston came close, to date no live platypus has ever set one of its flat feet on British soil, and maybe never will. Australia has been extremely guarded with certain members of its fauna since the 1920s, and has been largely unwilling to share them with the rest of the world. Right now, there are only two platypuses living anywhere outside of Australia. A female named Eve and a male named Birrarung arrived at the San Diego Zoo Safari Park in October 2019.[4]

In fact, in the last 63 million years, Australia and the USA are the only two countries where platypuses are known to have lived – and America only joined this elite club in the last century. As far as we know, since the platypus-like 64-million-year-old Argentinian fossil *Monotrematum sudamericanum* was alive, the only other occasions when there were living platypuses on dry land outside of Australia were when they were taken to America in 1922, 1947, 1958 and 2019. Each of these occasions required decisions from government to go against their 1921 ban on the export of platypuses (which was strengthened in 1933 to include any part of a platypus).

The arrival of the very first of those pioneering platypuses was thanks to Harry Burrell, who had trained animal dealer Ellis Stanley Joseph in how to keep them alive in captivity. Despite Burrell being the most expert mentor possible (he was the first to put platypuses on public display, in 1910 in Sydney), Joseph subsequently established a pretty poor record on that front – over the course of several years, many platypuses died in his care while he waited for the federal permits needed to leave the country with one. Eventually, in May 1922, he received the government's blessing, and set sail from Sydney with five male platypuses.

One died before they had left southern Australian waters. Two more succumbed from the shock of a massive wave crushing their tank not long after. A fourth died while the ship was at port in Hawaii. Joseph wrote:

My feelings can readily be imagined. I would rather have lost all of my shipment of a very valuable cargo of birds, animals, and reptiles. This was not because the platypus was worth more (far from it), but because it was my ambition to bring one alive to America. I am glad to say that good fortune eventually favoured me, since on June 30, 1922, I landed in San Francisco with the first living platypus ever brought to America.[5]

Or indeed to any other country. The rest of the journey was no less troublesome for the poor platypus. Constantly shaken by rocking and vibrations, he had to travel by train across the entire country, from San Francisco to New York, and Joseph barely managed to find enough worms to keep him alive. Nonetheless, 'On Thursday, July 14, [we] arrived in New York, both man and animal completely tired out.'[6] It's worth noting that Joseph was also carrying kangaroos, wombats and snakes on this voyage, among other species.

The Bronx Zoo's then director, William Temple Hornaday, was ecstatic. History had been made:

> The most wonderful of all living mammals has been carried alive from the insular confines of its far-too-distant native land, and introduced abroad. Through a combination of favouring circumstances it has been the good fortune of New York to give hospitality and appreciation to the first platypus that ever left Australia and landed alive on a foreign shore.[7]

On public exhibition for one hour a day, it lasted seven weeks before dying.

*

Over the course of recent history, certain individual zoo animals – particularly gorillas and pandas – have become popular celebrities in the public eye (and very often experience an equally celebrated afterlife, as museum specimens). In this fashion, in the late 1930s to '40s a pair of captive platypuses in Australia were to achieve worldwide fame.

Jack and Jill, two platypuses at Healesville Sanctuary, earned their stardom when they became the first to breed in captivity, under the attentive and enthusiastic care of David Fleay (who himself became a celebrity as a result). The unifying factor in seemingly every account of platypuses in captivity is the enormous challenge of keeping up with their appetite. They need a constant supply of live invertebrates, which has proved consistently hard to maintain. The cost of feeding Jack and Jill worried the sanctuary's management committee to the extent that they released a third platypus in their care back into the wild, despite the popularity of the species with the visiting public. Jill alone – who was a very small specimen – was recorded eating 400 worms, 338 beetle grubs and 38 crayfish in one sitting. She only weighed 900 grams (2 pounds), but her dinner that day amounted to nearly 800 grams.

The work to find enough live worms, crayfish and insect larvae to feed Jack and Jill's appetites was relentless, and it seeded an economy among enterprising locals who were paid to supply platypus food. One industrious Aboriginal family were able to buy themselves a four-wheel horse-drawn buggy with their earnings, and a local schoolboy bought a bike for himself and a milking cow for his family. They were paid sixpence per 500 grams (just over a pound), so these purchases represent a lot of worms.[8]

The pair were the zoo's main attractions. Its daily 'platypus show' could attract over 1,000 spectators, and the value to the local economy went beyond worm-catchers. The chairman of Victorian Railways train company recognised that the platypuses were generating significant business for him, as many visitors travelled to Healesville by rail, so he decided to help pay for their upkeep as a business investment.

In October 1943, after four years together – during a

platypus show, as it happens – Jill started to show signs that platypups might be on the cards, by collecting nesting material and taking it to a nesting burrow she had dug. Excitement grew.

At this time, pretty much everything that was known about platypus breeding behaviour came from Harry Burrell's lifetime of observation. Burrell had estimated that baby platypuses emerged from their burrows after six weeks (rather than the three to four months we now know to be the case). As that period came and went, Fleay grew anxious. The baby still hadn't been seen after nine weeks from when they estimated any eggs would have hatched. Although Jill was eating like a nursing mother, he couldn't cope with the uncertainty. Not only that, but a family of wild antechinuses – tiny carnivorous marsupials – had taken up residence in the platypusary, and Fleay was worried they might eat any babies or eggs. On 3 January 1944 he made the decision to dig into the burrow and see for himself. It was a massive risk, with so much to lose.

> Showing great anger at having her sacred nesting quarters raided, Jill began ejecting nesting material and soil in an attempt to block out the unwelcome intrusion of daylight. In the process she also pushed out a fat tiny baby who was still totally blind and helpless with only a covering of very short fur.[9]

Fleay was simultaneously horrified at what he and his team had done and elated with the discovery. This was the first ever platypus bred in captivity, but had they caused its death by digging it out? They hastily took some photos of the baby sitting in the palm of the hand, and then after several attempts to reintroduce it to the burrow while Jill continued to rebuff the intruders (including one point when she completely buried it in soil as she tried to build a blockage between her and them), they managed to return it safely to her.

Naturally, they worried about what Jill would do next. Had

they ruined everything? The wait must have been agonising. To their extreme relief, after a couple of days, her behaviour returned to how it had been before their incursion. They were satisfied that all was well and announced the momentous new arrival to the world press with great fanfare.

The story appeared in newspapers around the world – it was just the light relief people needed. Amid the horrors of page upon page of constant war reporting, on 25 March 1944 *The Illustrated London News* ran a full-page story: 'First Duck-Bill Platypus Born In Captivity'. Alongside a picture of the nine-week-old, one caption reads: 'Unaware of its zoological fame, this baby platypus has distinguished itself as the first of its genus bred in captivity. The duck-bill platypus, a kind of mole found only in Australia and Tasmania, has a bird's bill, is a mammal, and lays eggs.'*[10]

The text also includes the same old platypus slur: 'The platypus and the echidna (spiny ant-eater) are the last living descendants of the primitive reptilian creatures from which it has been contended all mammals have evolved.'

While I always appreciate a news story that points out that mammals are descended from reptile-like ancestors, I can't stop myself pointing out that by definition *all* modern mammals are the latest (but not last) living descendants of those 'primitive reptilian creatures', not just the monotremes.

As a result of all this media attention, visitors flocked to Healesville to see the platypus show. Jack had been holding the fort on public duty while Jill was nursing. Eventually, after seventeen weeks, the baby – a female they named Corrie – left the nest, and people loved her the world over.

Jack and Jill produced eggs again two years later – two of them – but they had to be removed, as feeding more than three platypuses would have been too much for the sanctuary

---

*I've just learned that in response to this press coverage some clever person wrote in to suggest that baby platypuses should be called platypups. So, sadly, this book is in fact not the first citable reference for platypups.

to cope with (the family of three required almost 2,000 worms and 60 crayfish every single day).[11]

Aside from an occasion in 1972 when an approximately fifty-day-old platypus was found dead in the entrance to its burrow at the wildlife park David Fleay subsequently established in Queensland, no one else managed to breed a platypus in captivity again for more than fifty years, until 1998, once again at Healesville. And since then, only Healesville and Taronga Zoo in Sydney have succeeded.[12] It is clearly difficult to replicate the circumstances for wild platypus courtship and reproduction in an artificial environment.

*

The brief appearance of the species in New York in 1922 clearly established a relationship between the city and the world's best animal. For in 1946, the Bronx Zoo once again requested the delivery of platypuses from Australia. And who better to oversee their risky export than 'the platypus man', David Fleay?

An expedition to transport any animal across an ocean is laden with challenges, but this truly was an adventure of epic proportions. Rosemary Fleay-Thomson tells the story in vivid detail in her father's biography, *Animals First*.

The first step was to find the right platypuses, and over a series of nights he caught nineteen, from which he chose the three he considered the most capable of crossing the Pacific. They were given the names Cecil, Betty Hutton and Penelope Platypus.

Precise plans for the platypuses' enclosure – with considerations for how to protect them from the New York winters – were sent over. The New York Zoological Society formed a Worm Committee to ensure the animals' dietary needs would be met, and agreed to pay to employ additional people in Victoria to meet the daily worm- and crayfish-demands of three hungry platypuses, which had temporarily doubled Healesville's platypus population.

When the time came, the ship MV *Pioneer Glen* was prepared for the journey. It would carry two concrete-lined travelling platypusaries, tons of fresh water, many crates of soil, 136,000 frozen worms, 23,000 live ones, 22,000 live beetle larvae, 7,000 crayfish, 45 live frogs, and preserved duck and hen eggs.[13]

But then, a potentially catastrophic telegram arrived. Despite having permits from the Department of Fisheries and Game, Fleay received a clear message from the Commonwealth Department of Customs:

MY MINISTER NOW NOT PREPARED APPROVE REQUEST AS EXPORTATION OF PLATYPUSES NOW NOT CONSIDERED TO BE IN NATIONAL INTEREST.[14]

Fleay had to throw every political ally he could muster at convincing the minister for trade and customs to reverse the decision. This involved the US ambassador to Australia, the Victorian state premier and a former prime minister of Australia. They argued – much like Churchill had – that this shouldn't be thought about as a commercial enterprise, where the beneficiaries were simply the Bronx Zoo and Healesville Sanctuary. Instead, the platypuses should be considered diplomatic envoys from Australia's Prime Minister Ben Chiefly to American President Harry Truman. They would be a show of gratitude for America's role in the war. This ploy worked, and the minister reversed his decision.

While these negotiations were taking place, the *Pioneer Glen* had already left Melbourne. Fortunately, Fleay had gambled on his trip to Canberra paying off and had instructed the crew to continue loading the platypuses' supplies and equipment before the ship departed, even though he and the platypuses were not on board. He returned to the Sanctuary and hired a hearse to transport the three platypuses, along with two echidnas, to the airport in Melbourne. From there they managed to fly via Sydney to Brisbane in the plane's noisy, unpressurised cargo hold and meet the ship before it left Australian waters.* In advance of their arrival, the local press had roused the people of Brisbane to take extra supplies of worms to the Queensland Museum, to be donated to the now-famous monotremes.

When they eventually left Australia, waved off by a crowd of journalists, Fleay felt the pressure of getting them across the ocean keenly. 'Should they die,' he wrote, 'following all the publicity, then only life as a refugee in Rio de Janeiro remained for me!'[15]

On the ship, disquieted by their exertions to date, the platypuses refused to eat frozen food – but there were not enough live worms and crayfish to last the journey. Fleay checked the route ahead and had a radio message sent to Pitcairn Island, which sits smack-bang in the middle of the South Pacific, halfway along a line drawn from New Zealand to Ecuador.

Pitcairn had been colonised by the British sailors who committed the infamous mutiny on the HMS *Bounty* in 1789. After leaving Tahiti, the *Bounty*'s crew set Lieutenant William Bligh and eighteen of his supporters adrift in an open, 7-metre-long (23-foot) boat in the middle of the Pacific with a compass, a sextant and five days' food and water. Bligh managed the extraordinary accomplishment of navigating his tiny vessel – which was so laden with men that its rim sat only 17 centimetres (7 inches) above the water – 5,800 kilometres

---

*These were not the world's first flying platypuses. In 1940, Fleay had flown ten from Essendon, near Melbourne, to Adelaide, on the way to introducing the species to Kangaroo Island. The descendants of these animals can still be found there today. Soon after, hearing of the success of the Kangaroo Island escapade, the Western Australian government wanted their own wild platypus population, and a single pair were flown from Victoria to Perth, but did not survive long.

(3,600 miles) through the dangers of the Great Barrier Reef, into and across the treacherous Torres Strait and all the way west to the island of Timor, arriving six weeks later.[16] He even managed to chart part of the northeastern Australian coast along the way.*

Upon seizing control of the *Bounty*, most of the mutineers returned to Tahiti, but their leader Fletcher Christian and eight others sailed away in search of a safe place to hide from any pursuit and court martial from the British Admiralty. Trapped aboard with them were twenty abducted Tahitians, most of whom were women. The crew thought Pitcairn was a good bet, as its location on maps was uncertain and so the navy would struggle to find them there. Indeed, it took Christian four months to locate it himself. They remained undiscovered until 1808, when an American sealing vessel stumbled across the island. The one surviving crewman – John Adams – was never punished.

And 139 years later, the 120 descendants of those mutineers and their Tahitian prisoners were asked to catch as many earthworms as they could and paddle out to meet the *Pioneer Glen* as it passed the island in the night, in order to feed three animals that they might never have heard of. The platypuses might not have survived had they not.

When they reached Panama – ready to pass through the canal and into the Atlantic – they were met with 10,000 more worms (known romantically in parts of America as 'night-crawlers', they are the same species as Europe's 'common earthworm') sent by the New York Zoological Society, and once again swarms of journalists to report on the journey so far – some of whom had flown over from Australia. Surpassing even that of Jack and Jill before them, few captive animals, if any, had ever attracted so much press attention.

---

*Bligh would go on to be the fourth governor of New South Wales. Despite his cartographic accomplishments, Bligh clearly lacked solid management skills, as not only did he suffer this mutiny, but he was deposed of his governorship by the military in 1808.

A little under a month after leaving their native shores, on 25 April 1947 (Anzac Day, as it happens – Australia's day of remembrance, one of the biggest days in the national calendar), Cecil, Betty Hutton and Penelope docked in Boston. They were driven to New York in a limousine.

And the crowds went wild. At a special ceremony at the zoo, Australia's ambassador to the USA formally presented the platypuses to the people of America on behalf of the Australian government, as a 'gesture of closeness and goodwill'.[17] Around 5,000 people turned up at the zoo on the opening day.[18] The press provided constant updates on the animals. With the Australian flag flying alongside the Stars and Stripes above the enclosure, if one of Australia's strategic goals for this diplomatic mission had been to raise public awareness of their fauna, then it was certainly a success.

Nonetheless, we can detect the familiar othering, 'weird and wonderful' tone in the many *New York Times* articles that appeared at the time (to pick just one publication). 'Some very strange animal life is found in Australia, but none stranger than the platypus', said one.[19] Others debated the animals' IQ.[20] An article written by David Fleay himself – the species' greatest advocate at the time – added to the impression that they were peculiar little curios:

> The arrival in New York of three duck-billed platypuses, those odd little Australian animals that represent some of the confusion of nature's efforts to turn out a mammal, has emphasized again that the big continent 'down under' possesses some of the queerest creatures on earth.[21]

The Australian Associated Press noted that the platypuses were 'receiving more publicity than any other Australian news item for months',[22] and the Australian ambassador considered the event to be the best publicity Australia had received overseas for many years. Clearly the customs telegram that claimed that the departure of the platypuses had not been in the national interest was mistaken.

Fleay spent months in America helping to settle the three amphibious diplomats into their new home and to train their keepers in their care, and his efforts made all the difference. Betty died around a year after their arrival, but Penelope and Cecil were New Yorkers for over a decade. They even exhibited some stages of breeding behaviour (their failure to succeed was blamed on the presence of the pesticide dichlorodiphenyltrichloroethane (DDT) in their food, having run off from farmlands into American watercourses). August 1957 was the beginning of the end for the pair, when Penelope managed to escape. She was never found, but one can't imagine a platypus would last too long in the wilds of the Bronx. Cecil died by the end of the same year. It would be tempting to romantically describe this as caused by heartbreak or loneliness, but let's remember that platypuses are solitary except for when they breed.

Straight away, New York wanted three more platypuses, and Fleay was called on one more time. In 1958 he crossed the Pacific again, but this time by air. Although the journey was far quicker, the noise, vibrations and customs barriers to importing the earthworms' soil and the platypuses' bedding straw made the journey no less stressful.[23]

Patty, Paul and Pamela – as the new trio were known – were welcomed with a similar fanfare, but overall this was a far less successful project, as the animals didn't survive for more than a year.

Following that, until the arrival of Eve and Birrarung in San Diego in 2019, Australia did not allow any other platypuses to leave the country. In a 1998 government report titled 'Commercial Utilisation of Australian Native Wildlife', they outlined the argument quite clearly:

The issue of exporting Australian wildlife to overseas zoos, and in particular the export of koalas and platypus, is complex and controversial. Because they are quintessentially 'Australian', there is a strong demand for these animals in overseas zoos . . . There is a strongly held view that these animals should be kept in Australia. . . . for reasons relating to tourism, wildlife and in particular 'icon' animals should be kept in Australia so that tourists would be forced to come here if they wanted to see our wildlife.[24]

Personally, I wonder how many people have ever said, for example, 'I was planning to go to India, but now that I've seen a tiger in a zoo I won't bother'. Nevertheless, if you want to see a platypus, or indeed many of Australia's other iconic species, you have to go to Australia (or for the time being at least, San Diego). Tourism is a major part of the Australian economy, and wildlife is one of their key selling points. For that reason, it could serve to be protectionist. When it came to the 2019 export, the Humane Society International Australia said, 'The fact remains that there is no conservation benefit in exporting platypus overseas.' The society thought that federal environmental officials should 'keep our beloved Australian icon at home where it belongs'.[25] However, San Diego Zoo insists that its platypuses have a valuable ambassadorial role in inspiring the millions of people who visit the zoo to learn about Australian wildlife and champion its conservation.

I do recognise that I myself am using the platypus as an ambassador for all Australian wildlife in this book. Deservedly, it is the cover star and principal protagonist, but its story on the world stage – acting as a diplomatic envoy to represent all of Australia's mammals – opens the door to introduce the other key characters.

3

Echidnas: The Other 'Primitives'

Early into my first trip to Tasmania, shortly before my first platypus encounter with Toby, we were driving across a friend's farm. Passing through the furthest paddock on her property, set among a lush forest of tree ferns and huge gums, I spotted a ruddy brown ball about the size of a loaf of bread bumbling across a field. It was an echidna, and I was ecstatic. Before I could communicate what was happening, I jumped out of the moving car, vaulted the fence and ran through the wet grass. My friends stopped the car and followed.

We sat down about 5 metres (16 feet) from the animal. My not-so-subtle approach had made it tuck itself into a ball of spines, but after a minute it uncurled and continued its hunt, sticking its long snout into the ground in search of ants.

After twenty minutes, by complete luck it had made its way right up to me. I have rarely been more excited than in that instant.

Different kinds of zoologists can adhere to different stereotypes. They are generally recognisable by their behaviour (as well as their clothing styles *cough* socks-and-sandals-and-palaeontologists *cough*). Herpers (herpetologists: those who study reptiles and amphibians), for example, tend to grab every animal they see with their hands, despite their group being the most dangerous (which is no doubt why they do it). Plenty could be said about birders, but one common habit

is strolling around the countryside in neatly laundered camou-
flage gear and well-displayed expensive binoculars making
psssshtt psssshtt psssshtt noises at bushes, which mimics a
general bird alarm call and thereby causes whichever tiny
brown species they want to tick off their list to zoom out
from hiding in startled panic, so they only get the briefest of
glimpses. Entomologists (insect people) tend to talk loudly,
as their targets are rarely scared off by human sounds. I'm
reliably informed that myrmecologists (a tribe of entomologists who study ants) regularly sample their study animals to
see what they taste like in case it helps with identification.
And mammalogists (we who study mammals) creep quietly
from bush to bush, moving in tight packs to avoid creating
more silhouettes than we need to, and would never conceive
of unnecessarily touching a wild mammal unless we were
temporarily trapping them in a survey. Except for when we're
deliberately catching them for scientific monitoring, we tend
to keep our distance.

'A Scene in Tasmania, with Characteristic Mammalia', by Johann Zwecker – a short-
beaked echidna appears with thylacines, bandicoots and wombats in Alfred Russel
Wallace's 1876 book on the distribution of animals. As is typical of historic echidna
illustrations (and taxidermy), its hindfeet have incorrectly been pointed forwards.[1]

And so, when a wild echidna – of its own will – ambled up to me and rammed its face under my leg, I didn't know quite what to do. It remains one of the most incredible wild-life experiences of my life.

*

The role of echidnas in the story of how the world considers Australian animals is just as significant as that of the platypus, so it's important we get to know them properly. Like their platypus relatives, echidnas can be described as an amalgam of familiar and unique features. They have a toothless, pointy snout like a cross between a bird and an anteater. They lay eggs like a reptile, have thick spines like a porcupine and a pouch like a marsupial. Their front feet resemble spades and their hind feet point backwards. They walk like no other creature on Earth, as if at the behest of someone who is learning how to operate a remote-controlled robot, stopping and starting and changing direction every couple of steps. The way their limbs move, held at 90 degrees from the body, is unique. In every respect, echidnas are delightful.

At rest, their spines point backwards, with those on their backs converging on the midline, where they cross over. This gives them a visually pleasing reverse centre-parting. In my experience, echidnas appear to have at least two defensive manoeuvres in which they put these spines to use, which are employed depending on the perceived threat level. Like platypuses, they are very easily spooked, and their initial response to the sound of a cracked twig, for example, is to immediately hunker down in the way I described previously, by tucking their vulnerable heads and legs under their spiny bodies. This manoeuvre can be taken a step further by rolling up into a ball like a hedgehog. This is achieved by a layer of muscle under their skin which, when contracted, works like a draw-string, pulling their spines down over their limbs and heads and shrinking all of their extremities in towards each other under their bellies.

They have a large pompom of spikes around their short tails, and a ring of long quills around the outer edge of their round bodies and up over their shoulders. When they cower down into these tucked-up balls, these quills effectively form a spiny skirt around the part of their body that is level with the ground, which is exactly where a predator would attempt to flip them over from, in order to get to their unprotected bellies. Let's call this spiky-ball strategy 'DEFCON 2'.

If they think the threat is more certain, and if the ground underfoot isn't too solid, they can implement a more serious defensive manoeuvre and go to 'DEFCON 1'. They do jazz hands with all four feet and sink vertically into the ground, leaving only their dense shield of thick spines exposed at the surface (they can go even further and descend below ground in this fashion). It is an amazing trick to observe. Their long claws drill through the soil like a hand-held blender and all of a sudden they have descended through the earth at rapid speed before your very eyes, with what seems like nothing more than a shimmy. Once in this position, they clamp themselves in place by locking their claws into the soil and any plant roots and pebbles around them. When ecologists are trying to temporarily catch an echidna as part of a study – to take samples or attach tracking devices, for instance – if the echidna fixes itself into the earth like this, it is effectively impossible to pick them up. Predators will find it equally difficult. Unless they dig out each foot from where it is clamped into the soil, the solid skeleton is arranged so compactly that the strength required to lever the animal out is beyond anything that is likely to eat them. A predator's – and an ecologist's – only real chance of getting hold of an echidna is to grab it before it realises there is a threat. They can also point their spines individually in different directions at will, so even if you do get them off the ground, they can still spike you.

As well as all that, they can also use their spines and under-lying muscles to wedge themselves into cracks in rocks or hollow logs. It's fair to say that echidnas seem to be winning in their arms race with predators, and as adults they are unlikely to form a significant portion of anything's diet. I have found Tasmanian devil poo containing echidna spines (ouch), but this is probably a result of scavenging dead echidnas, rather than overpowering live ones.

The arrangement of their reverse-oriented hind feet, which enables these protective strategies, is quite remarkable. One or sometimes two of the toes have elongated sickle-shaped claws, which at rest curve backwards and inwards towards the midline, lying flat against the ground. This also makes possible a further nifty technique. Because their legs are set in a sprawling position, rather than directly underneath the body, they can twist them up, around and over their backs, so that the long claws can be used to scratch and groom in between their spines almost anywhere on their bodies. Echidnas appear box-like and solid, so such flexibility is surprising. Watching them do this makes me feel slightly uncomfortable, since if any other animal pulled itself into that position it would surely involve some painful disloca-tions. Grooming is naturally difficult for spiny animals as their defences work equally well against their own hands and mouths as they do against those of predators. This is presumably why hedgehogs are famously riddled with fleas. The long, curved claws and reverse, sprawling orientation of their limbs is a neat adaptation for echidnas to tackle this problem.

*

As he would go on to do with the platypus in 1799, echidnas were first introduced to the Western world by George Shaw, in 1792: 'This extraordinary animal may well be considered among the most curious and interesting of quadrupeds yet discovered', he wrote (his subsequent description of the platypus was even more superlative).[2] What delighted and confused Shaw was that the echidna's characteristics were so familiar – just not in this combination. His description was rather like mine – he considered the tubular, toothless snout and long, wormlike tongue to be quite like an anteater's (which he assumed they matched in terms of their lifestyle), and the keratinous, hollow defensive spines to be like a porcupine's.

Shaw dithered about where they should be placed taxonomically, suggesting that a new genus may need to be created for them, but in the end decided to lump them in the same group – *Myrmecophaga* – as South America's giant anteaters. He gave them the common name 'porcupine ant-eater', and they have otherwise become known as the 'spiny anteater', 'Australian porcupine' and 'New Holland hedgehog'.*

Over the years they were subsequently put in the same genus as porcupines and then platypuses, but eventually taxonomists followed Shaw's original notion and gave them a name of their own. The one we have settled on is *Tachyglossus*, meaning 'swift tongue', with the species name *aculeatus*, or 'prickly'. This is perfectly descriptive – echidna tongues can flick out 100 times in a minute in search of their insect prey.

One of the strangest, most surprising and most fondly remembered social experiences of my life took place in the tiny township of Adventure Bay, on Bruny Island, off Tasmania's south coast. Toby and I were there for the island's incredible wildlife – particularly the abundant echidnas. We spent a couple of weeks sleeping in our car – a minuscule, ancient bright-yellow Holden Gemini – in the empty beach parking lot overlooking the Southern Ocean (it was, without exaggeration,

---

* The etymology of the word echidna is an interesting one, as *ékhidna* in Ancient Greek means 'viper'. The word *ekhînos*, meaning 'hedgehog' (or, if Latinised to *echinus*, also 'sea urchin'), would make much more sense, but that doesn't explain where the 'd' came from.

the only car in the state that anyone would rent to a foreigner who had just passed their driving test). It didn't matter though, as we were in an amazing spot: one morning we awoke to see a southern right whale swim past, penguins waddled in between us as we ate dinner and we saw more quolls out hunting at night than we could count. The boat ride we took with Rob Pennicott – who has since established one of Australia's most successful wildlife tourism companies – was genuinely awe-inspiring in a way that only nature at its best can be. We were surrounded by a megapod of hundreds of dolphins and fur seals playing and hunting around the boat while albatrosses wheeled above our heads, all with the backdrop of some of the highest cliffs in the southern hemisphere (upon which we would later find an echidna just a few metres from the top). The only downside to Bruny Island is that there are no platypuses.

Over the course of our stay we got to know the people who ran the Adventure Bay General Store quite well, as every morning we would wake up and the car battery was mysteriously flat. They refused to sell us jump-leads, instead generously insisting on lending them to us each day. We couldn't work out what was happening. Each night we would instigate an increasingly draconian rationing of electricity in an attempt to avoid draining the battery. No lights. No radio. No short journeys during the day. No keys in the ignition. But even then, in the morning: no power. On perhaps the fifth night, I woke up and noticed from the glow in the heavy condensation on the rear windshield that the car's brake lights were on. We hadn't realised that even with everything switched off, if Toby – in the driver's seat – was sleeping with his feet on the pedals, the brake light would come on and drain the battery overnight. Mystery solved.

Through our daily visits to the store to borrow the leads, we had been encouraged to come to the Adventure Bay New Year's Eve party at the village hall. 'Bring your own food and grog and music', we were told. When the night came, a storm was raging – full-on thunder and lightning – and we pushed open the doors to the hall and rushed in out of the torrential rain. It was like that scene in *An American Werewolf in London* when the two Americans enter The Slaughtered Lamb pub, and everyone suddenly stops talking and stares expressionless at the two outsiders at the door.

Perhaps naively, we had expected a reasonably sized gathering. Adventure Bay is one of three settlements of any size on South Bruny, and we assumed most of the 150 or so residents in reach of the village would be there. As it turned out, there were exactly twelve people inside. Twelve people and a wallaby. And they were all waltzing when we walked in (including the wallaby – no joke). We sheepishly introduced ourselves, which didn't take long (and in any case everybody recognised us as the two boys they had spotted sleeping in the car by the beach). Despite the arresting start, we couldn't have been more welcomed. Over the course of the evening we were taught the waltz (which they had been practising since February for this occasion – it was very *Strictly Ballroom*), and I spent much of the evening with the wallaby, which had been rescued from its mother's pouch when she had been killed by a car, it having been too young to leave the comfort of a warm body.

From then on, we would be woken in the mornings by locals knocking on the car window with warm muffins and the like – we felt like part of the family and I have been similarly welcomed when I've returned since. All of this is a roundabout way of saying that I am very fond of Adventure Bay and its many echidnas, and as it happens the first written record of an echidna comes from there. As eighteenth-century animal comparisons go, it's a good one.

On 9 February 1792, George Tobin, a lieutenant on board HMS *Providence*, under the command of William Bligh (yes, the same

Bligh we met earlier),* wrote in his journal, 'The only animals seen, were the Kangaroo, and a kind of sloth about the size of a roasting pig with a proboscis two or three inches in length . . . On the back were short quills like those of the Porcupine . . . this animal was roasted and found of a delicate flavour.'[3]

Before they ate this spiny 'sloth', Bligh also made some observations for the ship's log, including:

> Lieut. Guthrie in excursion today killed an animal of very odd form . . . it can scarcely be said to have a neck. It has no mouth like any other animal, but a kind of Duck Bill, 2 ins. long, which opens at the extremity, where it will not admit above the size of a small pistol ball. . . . It has no tail, but a rump not unlike that of a penguin, on which are some quills about an inch long.[4]

But despite the coincidence that this also happened in 1792, this cannot be the source of George Shaw's echidna description, when he formally named the species. Firstly, because the crew of the *Providence* ate their specimen, and secondly, because they wouldn't have got back to England in time for Shaw to get it before he published his account. Indeed, Shaw may never have seen a specimen himself, instead using a drawing attributed to the Port Jackson Painter from around 1791. It seems the first echidna specimen to reach Europe was preserved in a barrel aboard the Dutch ship *Waakzaamheid*, which left the settlement at Port Jackson in April 1791.[5] It was sent by the first Governor of New South Wales, Arthur Phillip, to the man whose name is written all over the history of Australian zoology, Joseph Banks.[6]

---

* Adventure Bay was known as a perfect spot for ships to stop as it is easy to anchor there and there is a ready supply of fresh water right by the beach. It gained its English name when the HMS *Adventure* landed there in 1773 while separated from Captain Cook's HMS *Resolution* on his second Pacific voyage. And Bligh himself had been there before, as captain of the HMS *Bounty* in 1788. It was the last place he stopped before Tahiti, the site of the famous mutiny.

For decades, the official count of living echidna species has been four, found in Australia and New Guinea, but how the different populations are related, and which actually represent separate species and sub-species, are questions currently being investigated by zoologists. At the time of writing, one species of short-beaked echidna and three species of long-beaked echidnas are recognised. However, one of them – Attenborough's long-beaked echidna* – is only known from a single specimen collected in 1961 and has not been seen since. For many years the suspicion has been that it is not sufficiently different to warrant being called a full species, but we will have to wait for the results of the current studies to come out before definitely saying what the true number of echidnas is.

Being one of Australia's most widespread mammals, most of the details we know about echidnas' lives come from short-beaked echidnas. The long-beaked echidnas, on the other hand, are critically endangered and found only in areas of New Guinea that are hard to access. As such, unless I specify otherwise, the descriptions in this chapter refer to the familiar short-beaked animals. To give a sense of what they look like,

A western long-beaked echidna, illustrated by the Spanish naturalist
Angel Cabrera in 1919.[7]

---

* Named for the great man.

long-beaked echidnas can reach more than three times the length of a short-beaked – a real whopper could be a metre (3 feet) in length and weigh as much as 16 kilograms (35 pounds), but 60 centimetres (2 feet) is more typical. As the name suggests, their snouts are a lot longer – 12 centimetres (5 inches), rather than 7.5 in the short-beakeds – and also somewhat downwardly curved.

Short-beaked echidnas are found in every corner of Australia (as well as parts of New Guinea), from the Snowy Mountains to monsoonal savannah, from desert to rainforest, from eucalypt woodland to vine thickets, from coastal heathland to arid-zone claypans, and from sandstone escarpments to towering pine forests. Even though they are restricted to that one corner of the world, short-beaked echidnas have proved highly adaptable. They are sensitive to overheating, so if it's hot they forage at night. In regions that become icy over winter, they hibernate.

From the fossil record, we know that some species of long-beaked echidnas lived in Australia during the Pleistocene Epoch, which ended at the close of the last Ice Age, around 12,000 years ago (Australia and New Guinea are connected when the sea level drops during Ice Ages). There is also Aboriginal rock art in Arnhem Land in the Northern Territory that seems to depict long-beaked echidnas and which appears to date from the Pleistocene. As such, though they were once present, long-beaked echidnas were not considered part of the modern Australian fauna.

However, in 2009, a friend of mine, Kristofer Helgen – who has a knack for making new mammal discoveries from museum collections – was studying the echidna specimens in the care of another friend, Roberto Portela Miguez, who curates the mammal collection at London's Natural History Museum, when he found something extremely surprising. Attached to the leg of a long-beaked echidna skin were two labels, which listed the place that it was collected as 'Mt Anderson (W Kimberley)' and the year as 1901.[8]

The Kimberley region is an area in the tropics of north-western Australia, and Mount Anderson is a hill* 90 kilometres (56 miles) southeast of Derby, the region's main western gateway. Kris and Roberto have relayed the moment of the discovery to me on a number of occasions and they were certainly excited. This would mean that long-beaked echidnas lived in Australia not only beyond the end of the Pleistocene, but into the last century.

As seasoned museum professionals, the two knew not to take information on museum labels at face value. Museums get things wrong quite often. Species names change, labels get switched, specimens are incorrectly identified – and curators are only human. So, Kris and Roberto followed the paper trail from the original notes of naturalist John Tunney, who had purportedly collected the specimen. It all added up – there could be little doubt that the specimen is exactly what it appears to be.

Their discovery was huge (at least for Australian mammal fans). Even if it hadn't been seen since 1901, Australia's list of modern native mammal species had just got one echidna longer. To add to the excitement, one of their colleagues, James Kohen, recalled a 2001 conversation with an Aboriginal woman from the northern Kimberley who had suggested that her grandmother had experience of two species of local echidna, one far larger than the short-beaked.

Tunney's original notes suggested that the specimen was caught in rocky country. Although long-beaked echidnas in New Guinea are most associated with tropical rainforest, they do make use of other habitats, and Mount Anderson isn't inconsistent. However, not far north from there, there are little pockets of tropical rainforest and vine thickets that one might better associate with the species.

---

* Certain parts of Australia can get a bit overexcited when deciding what gets called a mountain – this one is less than 300 metres (1,000 feet) high.

Tunney's was the first mammal survey of the region, which has never been densely populated by European or Indigenous Australians. A year before Kris and Roberto published their paper on the long-beaked echidna specimen, I began visiting the Kimberley regularly to assist with the ecological fieldwork being done there by the Australian Wildlife Conservancy (AWC). It's fair to say that since Tunney, faunal surveys have been fairly localised in the area, and the Kimberley is both vast in size and diverse in the range of ecosystems it harbours. The region has had a number of new mammal discoveries in recent decades, and it is well within the realms of possibility that the long-beaked echidna hangs on in a remote pocket of the Kimberley today. Whenever I am out there – particularly in the wetter regions of the north – I constantly dream of meeting Australia's lost echidna.

*

If you come across an echidna and it becomes so startled that it goes to DEFCON 1 and buries itself, then unfortunately it may not resurface rapidly. I have sat silent and motionless with a self-interred echidna for almost an hour without it making a move. However, if it only curls itself up, and remains at DEFCON 2, before long it might gingerly untuck itself, sniff and look for danger. If it decides that it is unlikely to be attacked, it will resume normal echidna business. This means that you can watch echidnas forage at very close quarters.

On one such occasion, I had been sitting a couple of metres away from a busy individual for an hour or so when it struck me that through the mind-blowing marvels of biology, nature had converted millions of ants into something so brilliant as an echidna.*

---

* I know that's how food works for all animals. But still . . .

Echidnas forage in a seemingly random way. They will walk a few steps and thrust their nose into the soil or leaf litter and appear to take a little sniff. If something catches their interest, they will ram their face even further down in search of their prey. Although the bones at the front of their skull seem fragile, they can lever rocks out of the way, break open rotten logs and delve into soil with their 'beaks'. I've watched them appear to throw all the strength they can muster into forcing their snouts into the ground, with every muscle straining in their upper body.

That first sniff suggests that smell is important in finding food, and this is backed up by the structure of their brains. Despite the fact that – just like platypuses – echidnas are consistently described as 'primitive', they have extraordinarily large brains for animals of their size. This is somewhat surprising for a species that is considered to be largely solitary, and suggests that their social relationships, over the course of their long lives, may be quite complex.[9] In any case, the part of their brain that receives information from their smell receptors – the olfactory bulb – is particularly significant in echidnas: they are the only mammals known to have folded olfactory bulbs.[10] Folding in the brain (all the grooves and wrinkles in a human brain, for example) is a means of increasing surface area, and thereby the region's functions. Broadly speaking, the more folded a region of a brain is, the more effective it is. Echidna brains are very good at smell. Aside from feeding, it is also assumed that detecting scents is important in communication between echidnas, particularly around mating.

In addition, echidna snouts are equipped with touch sensors, and – like their platypus cousins – receptors that can detect electricity (though the echidnas' are far less sensitive than platypuses'). Electro-reception can only work in water, so why do they have this impressive adaptation when they live on land, including some of the driest habitats on Earth? There is evidence that echidnas, despite looking radically different, evolved from a swimming platypus-like ancestor, possibly as recently as 15 million years ago.*[11] Having inherited this trait from them, echidnas continue to make use of it when they can. The most likely explanation is that when the soil is wet enough to conduct electricity, electro-reception proves to be a handy additional tool for finding food, but they don't totally rely on it. This is supported by the fact that the long-beaked echidnas of the tropics (where the soil is wetter) have far more of these sensors than the short-beaked echidnas that live across Australia. The skin of echidna snouts is also moist, so presumably they can pick up signals if they come into direct contact with prey. To add to this intrigue, while I was writing this, Stewart Nicol (a leading echidna researcher at the University of Tasmania) sent me a video of an echidna foraging underwater. Its movements looked exactly like a foraging echidna on dry ground (except for occasionally raising its head to breathe), but it was submerged. Who knows what information it was receiving from its electro-receptors?

Echidnas are actually very capable swimmers – using their snout as a snorkel – which again lends weight to the idea that they evolved from platypuses. Primarily, however, this is thought of as being a tactic to cool down in high temperatures. Unlike platypuses, which are freshwater specialists, echidnas have been found swimming out into the ocean.

We already know that echidna bodies are built for impressive digging abilities, and these are crucial for their feeding habits. Their preferred food is termites, which are far softer than ants, but they'll eat whichever are prevalent in their area (and will mainly eat the species without unpleasant defence

---

* This date is by no means universally accepted. The fossil record for echidnas is extremely poor, because they have thin skulls and no teeth. The oldest fossils are only 15 million years old. Without many fossils, it is very difficult to accurately assess the rates of genetic change, which are used to estimate the dates when two groups split in their evolutionary history. In 2021, one major study of platypus and echidna genomes suggested that they split from each other as long as 55 million years ago (Zhou et al., 2021).

mechanisms). Both of these groups are social insects that can be found in extraordinary numbers within their nests. Echidna forelimbs have powerful muscles and broad, strong claws for breaking into these structures. In certain parts of Australia – particularly the monsoonal woodlands of the north – termites build their nests by cementing soil particles together to withstand submersion in floodwaters. As such, the constructions are rock hard and can reach metres high and wide, but echidnas can break their way in. If termites and ants are not available in sufficient quantities, echidnas will also eat beetle larvae and other invertebrates.

Once they have found what they are searching for, they will rapidly flick their tongue in and out. An echidna tongue can reach 18 centimetres (7 inches) beyond its mouth (i.e. more than half their body length) and is covered in gluey saliva so that insect adults, eggs and larvae in the nest all stick to it. The back of the tongue has horny pads on its surface, and as it is retracted into the mouth, the food is mushed up against corresponding pads on the palate.

This mode of relatively indiscriminate licking means that they end up eating a lot of soil and nest debris, which gives echidnas very distinctive droppings. They come out in long, dense cylindrical tubes, like a cigar made entirely from fine-grained earth. When you break them open, they reveal thousands of crushed insect exoskeletons, particularly when found in areas where the echidnas chiefly target ants. The droppings are often left in latrine spots – one or many echidnas poo consistently in exactly the same place, leaving strong scents that may signal their home ranges to other echidnas.

At the other end, their mouths are absolutely tiny – just 5 millimetres ($^1/5$ inch) wide at the tip of a beak that is 6 centimetres (2 inches) long. If they look straight up into the air, their snouts resemble the strongly downturned mouth of Beaker from the Muppets – a sharp, upside-down U shape.

Inside their mouths, their lower jaws have been reduced to

thin, slightly curved rods just a couple of millimetres thick. These probably don't have a significant role in chewing – which is done entirely by the tongue – since the muscles are minute, but, uniquely, the jawbones open the miniature mouth by rotating along their length so the tongue can extend out. These bones may also be used in the transmission of sound from the earth to the ears, meaning that they listen with their jaws for prey in the soil. This reflects where the mammal ear bones come from, evolutionarily: the bones at the back of a reptile's skull correspond to the bones inside the ears of mammals. Snakes also listen with their jaws.

Aside from pillaging social insect nests, echidnas will simply dig into soil to find their food, and I've watched them follow their senses diagonally downwards into the earth until they have pushed so hard and so far that their entire bodies are rammed below the surface in pursuit of some morsel. People lucky enough to have echidnas visit their garden will soon know about it, as they will end up with conical nose-poke holes all over their lawns, and then occasional deep furrows that are the full width of an echidna. As they hunt, a single echidna can turn over more than 200 cubic metres (7,000 cubic feet) of soil a year, meaning that despite their small size they play a vital role as ecosystem engineers by aerating soils across the Australian landscape.[12]

As well as being sticky, the tips of their tongues can also flex to grip larger prey items. If an item is too big to fit in their little mouths, like a beetle or grub, they will crush it under their beaks and lick up the soft bits. While long-beaked echidnas do eat social insects, as well as centipedes and crickets, they are believed to be more specialised for eating earthworms – particularly in wetter habitats – and have a grooved tongue with three rows of horny, backward-pointing spines to help grip these slithery invertebrates.

The universal occurrence of ants and termites across all Australian environments is presumably a major factor in why echidnas have such a widespread distribution across the country. It's quite surprising that, given their availability, more Australian mammals don't rely on these insects. Numbats, the stripy, squirrel-sized marsupials now restricted to the very southwestern corner of Western Australia, are the only other species to do so. Perhaps the success of echidnas has prevented other species from evolving to fill this niche.

Despite the abundance of their food, it is low in energy, and echidnas have a low metabolism as a result – possibly the lowest of any mammal.[13] This comes with a very long – but relatively slow – life, of perhaps thirty or fifty years, and they don't start breeding until they are five or six. Like pretty much everything to do with monotremes, echidna reproduction is a fascinating story.

*

An old, old joke about spiky animals goes,

'How do hedgehogs mate?'

'Very carefully,' is the answer. In the early 1900s, Harry Burrell put some thought into this question with regards to echidnas, and his conclusion was that echidnas mate face to face – and the easiest way for them to do this is if they stand up on their hindlegs, belly to belly, and support each other 'until the desired position is attained'.

There aren't many mammals that mate face to face like humans can, but Burrell was led to believe that this was 'the desired position' for echidnas because his field assistant, Bill Lancaster, said he came across mating echidnas on a mountainside. The account shows that if the echidnas were doing it 'very carefully', Mr Lancaster certainly was not exhibiting the same level of consideration:

> He described the participators as forming one large ball of
> quills with a pair of muzzles just protruding at one end, the

whole resembling a spiny melon with split stalk attached. After a minute or two Lancaster rolled them over with his boot and, as this had no effect, he deliberately kicked them down the mountain side, but even this failed to separate them and they remained together for some time after reaching the flat ground below.[14]

Based on this story alone, Lancaster doesn't sound like too solid a fellow and perhaps Burrell would have been wise to doubt him. His account does not match our current understanding of how echidnas have sex.

If the female is receptive, she will lie flat on her belly. The act itself then involves the male digging from beside her to clear a space under her tail in order to allow himself room to position his lower body beneath her. This is achieved by climbing the front of his body over the back of her body from behind, and then adopting a sitting position behind her, possibly steadying himself with his forelegs on her back, lifting her tail with his hindlegs and bringing their cloacas together below her body. They both face in the same direction, with her tail pointing at his belly.[15]

Puggle-production is not always quite that simple, however, as during the breeding season females can attract multiple males at once, who follow her around single file, more or less nose to tail, in what are known as 'echidna trains'. In some regions, more than ten male echidnas may form a train, and the convoy can last for well over a month. When the female finally becomes receptive, several males may attempt to dig below her and push the other males out of the way. Females are likely to mate with several males to increase their chances of reproductive success.[16] As a result, male echidnas have large testes – because this mating system involves what is known as 'sperm competition'. There is evolutionary pressure on the males to increase their chances of being the one to father the single egg that the female lays. This can be achieved before mating has taken place, for example by fighting other males off; or during, for example by increasing the volume of sperm he produces. In addition, it has been found that echidna sperm move in 'bundles', which is believed to increase the efficiency of their swimming as they race to reach the egg, a bit like a cycling peloton.[17]

Echidnas are obviously sufficiently defensively equipped for the females to be in control of when mating occurs. Or so you would think. In areas where echidnas hibernate, such as Tasmania, the males can awake from hibernation far earlier than the females. After feeding himself up, a male may find a still-hibernating female and mate with her before she properly 'wakes up'. He may then stay with her for several days to ensure no other males come along and do the same thing, before leaving to find more females. It's one of those biological stories that your human sensibilities can't stop you from counting against the greatness of echidnas.

The echidna penis has four heads, like four spiny footballs on the end of a fat pole. Each head has a rosette of append-ages – they have been described as resembling four little sea anemones. I've not come across an explanation of the purpose of these rosettes, but I wonder if they are also an outcome of sperm competition – could their role be to scrape out any previous males' sperm from the female's reproductive tract? This is one theory for why the human penis is shaped the way it is, with a distinct head. It is rather remarkable that it took until 2021, 229 years after the first scientific account of the species, for a full anatomical description of echidna penises to be produced, but some questions remain unanswered.[18]

For some time, a notable mystery surrounded the echidna penis: every time researchers encouraged one to become erect (through electrical stimulation, in case you were wondering), it engorged to a size that was significantly larger than a female's reproductive tract. How could such an organ be put to use? This question was only solved when a zoo animal had to be retired (by donation to a research lab) because he got an erection every time he was handled by the public. Through this excitable individual, they discovered that when it happens naturally – rather than being zapped with a mini Taser – the penis can fit into a female's tract because only half of it becomes erect each time, and they alternate sides from encounter to encounter. Nevertheless, when erect, their penises are 7 centimetres (3 inches) long and over a centimetre (⅓ inch) wide, which is equivalent to one-fifth of their entire body length.[19]

The egg-laying abilities of echidnas were as controversial as those of platypuses – as we shall explore in the next chapter – but we know now that about twenty-one days later (or as many days longer than that as the female remained in hibernation after she became pregnant, in those scenarios), a single egg is laid, or very rarely two. This is immediately placed in

A female echidna on its back, from Richard Owen's 1865 account of echidna pouches and young. The pouch is shown with one puggle nestled within. The huge digging fore limbs, backwards-pointing hind limbs, and the opening of the cloaca can also be seen.[20]

the muscular pouch that develops in the female's abdomen during the breeding season. The egg hatches after around ten days, when the puggle (as the babies are known, remember) is around 1.5 centimetres (½ inch) long, and weighs just half a gram. Nevertheless, it will have well-developed forelegs, with claws that allow it to grip onto hairs in the pouch.

After about seven weeks, the puggle will have outgrown the pouch and started to develop spines. At this point, the female leaves it behind in the burrow while she heads out to forage. As time goes on in the puggle's growth, the mother may leave the infant alone for as long as five or six days before returning to nurse it. And even then, she may only stay with it for a couple of hours at a time. The age at which they are weaned varies significantly across their range: in Tasmania it's four to five months, while on Kangaroo Island in South Australia it's nearer seven, but overall they tend to reach the same weight of over 1 kilogram (2 pounds) before they are left to fend for themselves.[21]

With their conical heads, significant claws and naked, pink, rotund bodies covered with deep wrinkles and dimples, puggles are some of the most remarkable baby animals out there. With all this considered, it's awful, but not surprising, that a black market has developed for pet echidnas.* They are extremely difficult to breed in captivity, so when echidnas were advertised in Southeast Asia as being 'captive-bred', suspicions were aroused. To uncover the fraud, scientists at the Australian Museum and Taronga Zoo in Sydney developed DNA tests and chemical analysis of their quills to establish whether the echidnas on offer had in fact been poached from the wild.[22] I was disappointed that the ensuing newspaper headlines didn't read, 'Smuggled Puggles'.

*

---

* Please, please, please don't ever entertain the idea of keeping a wild animal as a pet.

Animals move in very particular ways, and this allows us to identify them from the sounds they make as they move around. I have often been woken in the night by the noise of an animal moving through leaf litter just outside my tent. Being able to recognise what animal it might be from the distinct characteristics of the crunching leaves is worthwhile, as it allows you to decide whether it would be sensible to crawl out and take a look. For example, when I have heard the soft padding of a wolf, the hectic swish-swish-swishing of a herd of wild boar (something I experienced during what rates as one of the worst and most terrifying nights of my life*), the cushioned gallop of a hyena, or the rhythmic, gentle-yet-significant plod of an elephant, I felt the wisest move was to stay put in my sleeping bag. As tempting as it might be to stick your head out and take a peek, startling these species at close quarters while all you have protecting you is a sheet of tent nylon would be risky.

Australia arguably has some of the easiest mammals to identify in this way. Kangaroos and wallabies are obvious because of the unmistakable sound of a hop, and its volume and frequency should give you an indication of size. The hesitant slap/thud of a massive introduced cane toad – like a water balloon hitting a wall but not bursting – is all too familiar in the tropical north where they continue to spread like a plague. The snowplough shuffle of a wombat is also pretty distinctive, and the sharp rustle-burst of a bandicoot only slightly trickier. One on occasion I heard gliding possums softly crashing from sand palm to sand palm above me and was about to get out and check the species until one glided (glid? glade? glud?) into the side of my tent. This identification ability can be put to good use not just when you are trying to sleep, but also when you are actively searching for animals.

---

* They were out there for hours as we had camped right by the only spring in the whole valley – it seemed like a good idea to have water on hand, until it was apparent that every other animal in the area thought so too.

But sound is not the only guide. Since nearly all Australian mammals are largely nocturnal or crepuscular (active at dawn and dusk), your best bet for finding many of them is to walk around at night. Torches are pretty decent these days, and in recent years we've gone from having to carry a weighty car battery in our backpacks wired up to a massive hand-held spotlight to being able to see a fair distance with a high-powered head torch.

When scanning a torch beam around for signs of life, often the first thing to catch a mammal-watcher's attention is eyeshine. Many nocturnal (and deep-sea) animals have a structure in the light-sensitive layer of their eyes called a *tapetum lucidum*. This improves their ability to see in the dark by reflecting the little light there is back across the layer of retinal cells that can detect it, effectively giving a ray of light a second chance of being picked up. When these animals are caught in a narrow beam of torchlight, their *tapeta lucida* reflect the beam back with incredible brightness, and two vivid spots suddenly gleam in the darkness. Even when you are looking well beyond the point at which you can make out any shapes in the darkness, this eyeshine tells you an animal is there as obviously as a neon arrow. Depending on the species, eyeshine can be visible from several hundred metres away, and for crocodiles, fishes and frogs it even works when the animal is underwater. This feature of nocturnal animals inspired the invention of 'cat's eyes' in roads.

For some species, the only real hope of ever finding them is from their eyeshine. For example, when surveying for geckos, whose bodies are camouflaged against rocks or trees and who don't move until you are very close (when eyeshine doesn't work anyway), you can only spot them from the bright dots reflecting in their eyes. There are some pitfalls, as spiders and moths have extremely bright eyeshine, and often sit in the same places as you might expect to find a gecko, leading to possible confusion. However, insect and spider eyes twinkle in the torchlight as it moves across their different lenses, or different eyes (most spiders have eight). And I can't tell you how often a star shining through the branches has caught my eye when I've been out searching for possums.

The main drawback of eyeshine is that it only works if the animal is looking at you (and has its eyes open), and it is also very hard to use this method in the rain or fog. Plus, not all animals have a *tapetum lucidum* (most primates included). In the end, the sound of an animal moving through the undergrowth is always a huge giveaway that there is something out there in the darkness.

The Kimberley is one of the most exciting places I have searched for animals at night, thanks to the incredible diversity of mammal species found there. Some of these are restricted to tiny pockets of specific habitat, but as surveying efforts increase in the region, new populations are being uncovered. It is one of my absolute favourite places on Earth – possibly only surpassed by Tasmania (there are no platypuses in the Kimberley*). Since 2010 I have been lending my ecological expertise to the Australian Wildlife Conservancy on a voluntary basis, to help with its mammal surveys. The AWC is a brilliant environmental NGO that uses hard scientific data to learn the most effective ways of managing huge tracts of Australia for biodiversity. On one visit in 2011, between field surveys, I was at the main base camp identifying species from remote camera-trap photos taken at a location a couple of hundred kilometres away in the wetter northern Kimberley. I couldn't have been more excited to come across a scaly-tailed possum walking through one of the images.

Scaly-tailed possums, or wyulda, are a rare and declining species that today are mainly found in areas of the Kimberley

---

* Although with that said, I have seen rock art there that looks exactly like a platypus, so who knows?

with high rainfall and complex, broken rock piles where they shelter, with access to lots of different types of habitat nearby to feed in. They have long tails (covered in scale-like rough patches, presumably for grip) that are prehensile, so they can grasp branches like a fifth limb. Seeing them on camera was so tantalising – they went straight to the top of my must-see list so I could understand them better.

When my friend Rosie Hohnen later went on to do her PhD on this population, she spent nine solid months a year radio-tracking them across the escarpments, completely isolated, camping under a tarp in the middle of the bush with a very occasional helicopter food-drop. I was desperate to join her in the field, but I couldn't make the logistics work.

Instead, over the course of a few years I went on shorter trips that sought to establish how far the population ranged from the central 'core' that Rosie was studying. That meant going to places that wyulda were not known to live and looking for them and the other endemic species that they live alongside, such as golden-backed tree-rats, monjons (miniature, foot-long rock-wallabies), Kimberley rock-rats and golden bandicoots. A helicopter would drop us off with our equipment in some of the most remote locations in Australia and move us to a new site every few days or so. Every night, we baited and checked treadle traps; and during the days we set up cameras that would stay out for a few months after we left. These places had either never been surveyed by a biologist or were last investigated many decades earlier. Going on these expeditions is always an extraordinary privilege and an unbelievable experience, but for years, on each trip I took part in, no wyulda turned up in our traps. Once, an individual appeared on one of our cameras several weeks after we left, which was good news in that we had found a new site where they lived, but also somewhat frustrating since they had been right there with us but we hadn't seen them ourselves. It took me four years before I actually found a population in person.

Some time before that happy day, in the wet season of January and February 2014, I was in the Kimberley for one of these field trips. Starting at the AWC's main base in the Central Kimberley bioregion, I spent a week going through around 10,000 camera trap photos to help inform which sites we should visit to look for wyulda and the other specialist mammals. Combined with this information, the team scoured Google Earth to direct the decision, and the five of us eventually set off. To save on aviation fuel, the plan was to drive the team and the gear as far north as the dirt tracks would allow us, and then the helicopter would take us from there.

Fieldwork in remote locations is logistically complex at the best of times, but in the wet season the level of uncertainty caused by the weather and the situation on the ground means that you have to adapt quickly to changing conditions. A couple of hours after setting out from base, a huge monsoonal storm hit. The tracks turned into streams a few inches deep, and the soil beneath offered little traction. The rain was so heavy that it was like being in a carwash, but we made decent distance – until we were finally stopped when we came to a ford in a river that had become too deep to cross.

The rain cleared away in the early hours, and at first light the tiny helicopter arrived to start the complex flight plan of shifting us and our hundreds of kilos of trapping and camping gear to our first site (on one of those journeys we dropped off a supply delivery to Rosie's camp – I hadn't seen her for over a year, and it was bizarre to jump out of a chopper, give her a hug and a crate of tins and rice and fly off less than a minute later).

Over the next few days at our first site, we caught quolls, rock-wallabies and several species of rodent, but none of our target species. This was disheartening as it was the closest we would be to the known population of wyulda – and we had no more success at the next two sites.

Our final site was a stunning broad gorge further south.

The river flowed on one side of the gorge, and we set up camp in the dry, sandy bed of the other side, perhaps 200 metres (650 feet) from the water. We checked around for flood debris and were confident that it was safe as the water hadn't been on this side of the channel for years.

One of our lines of traps was along a beautiful little creek running through a vine thicket. Although we were further south than wyulda were known to live, it felt good. However, aside from this little pocket, the habitats in the vicinity were drier – mainly eucalypts and cypress pines rather than rainforest species. The rain was more predictable here, too – calm in the mornings but building up to incredible thunderstorms in the afternoons and evenings. Some nights, hours would pass when lightning flashed twice every second. And each morning, the river was a little wider and a little faster, and swimming across became harder and harder.

In the traps and while out spotlighting (when breaks in the rain allowed), we were just seeing rock-rats and rock-wallabies. I mean these animals no disrespect – it's always a pleasure to see them. Kimberley rock-rats are dumpy little things with excellent antipredator strategies. When something grabs them, their fur detaches in fluffy clumps, so the predator loses its grip. And if that fails, they can drop their tails like many lizards do in order to distract a predator from their bodies (there's an unpleasant in-between option, too – if something grabs their tail they can detach the skin only, so it slides off like a sock, leaving the naked tail bones and muscle below). They seem much slower than common rock-rats, which are also found alongside them, and this may explain why many of the Kimberley rock-rats we find are tailless. As I say, they are wonderful animals, but it can be disappointing to find them, as every trap that has a rock-rat in it can't catch a wyulda, since the first animal in closes the door behind it.

The morning before we were due to leave, we called into

base on the satellite phone to confirm plans, and were told that because of the weather, the itinerary had changed, and we would spend one more night out.

The little creek with the vine thicket ran into the main gorge on our 'dry' side, further downstream from our camp. It, too, had been running higher with each day's rain. We had been keeping a close eye on the height of the river, which was separated from the side of the gorge where we camped by a large, protective sandbank. We were confident that it was not anywhere near close to breaching.

However, a few hundred metres further down, the gorge narrowed to about 40 metres (130 feet) wide, with high-sided cliffs. At this point the creek below our camp was pumping more water into the river than could fit through the tight part of the gorge. Rather than breaching the sandbank from a mass of water coming down from above us, it began to back up from the constriction in the gorge downstream: the river started to run backwards up our dry side of the riverbed, gently filling like a bathtub with a clogged drain. Very quickly, we were under a few inches of slow-moving water.

Fortunately, we had been packing most of the day, and could easily stow our equipment higher up the sides of the gorge and were in no danger. As we sat on little boulders under our tarp, we were joined by all the spiders, scorpions and giant centipedes as they also sought refuge on these shrinking rock-islands. Eventually we had to retreat up the hill. By the time the storm had passed, the vine-thicket creek was churning, and the river was immense.

Despite all this, the change in plans had granted us one bonus night of spotlighting, and we weren't going to waste it. I went up near the top of the slope to a likely-looking rock castle – as we called the wyulda's habitat of complex blocks of sandstone separated by deep, narrow cracks where they like to shelter. On three occasions my heart leapt at seeing appropriately sized and coloured eyeshine for a wyulda up in the trees, but each one turned out to be a glider. It feels odd to be disappointed at seeing something so wonderful as a glider – they are beautiful, 20-centimetre-long (8-inch) possums, with striped faces and parachutes of skin between their limbs, which they use for gliding between trees. Seeing them in the Kimberley should add an extra level of excitement, too, as preliminary genetic studies suggest that the individuals found there may constitute an as-yet-undescribed species.*[23] However, it was not the animal I was hoping so desperately to see.

But then, I heard something on the other side of a rock pile. A sudden, heavy swish of the leaf litter. A pause, and then another. It was an animal gracelessly shoving its way through the layer of dead leaves and sticks, but haltingly. In the pitch black of a remote tropical forest, I had no doubt that this was the sound of an echidna.

.Naturally, as soon as Kris and Roberto had discovered that 1901 record of a long-beaked echidna, I had been urging my friends in the Kimberley to keep their eyes peeled. That individual was found only 200 kilometres (125 miles) from where I was. The reality was that in the five years that my friends had lived there, surveying the wildlife more intensively than anyone ever had in that region, they had only ever seen three echidnas, all short-beaked. But their base is in the drier central Kimberley ('dry' is a relative term in the tropics). The more time we spent in the wetter regions, the more we would

---

*In 2020, a new study added more intrigue to this assertion. The animals that had been called sugar gliders (which I had thought I had seen across Australia) were split into three species. The study stated that animals from the north should now be called savanna gliders (and are actually more closely related to squirrel gliders and mahogany gliders than they are to true sugar gliders); most animals from the east and Tasmania (where they are introduced) should be called Krefft's gliders; and sugar gliders are actually restricted to a small strip of coast in the southeast (Cremona et al., 2021). This would mean that I have never actually seen a sugar glider. On top of that, the study found that the newly named savanna glider is found in the Kimberley, but that some Kimberley specimens were different, and so may represent yet another new species. I'll never know which species the animals I saw that night actually were.

fantasise about finding a live long-beaked echidna. We knew it really could be out there.

I am always excited to come across an echidna, but when I heard that distinctive leaf-rustle of theirs that night, there in that wet, rarely visited part of the Kimberley, my heart was racing. On the other side of that boulder, was I about to confirm that Australia was still home to long-beaked echidnas?

The short story is no.

However, the animal I found did surprise me. It was a short-beaked echidna, but not as I knew it. The amount of hair echidnas have varies depending on where they are in Australia. Those in Tasmania, for example, where it gets cold, are so hairy that the spines on their back are almost obscured. I have seen hundreds of echidna specimens in museum storerooms, and so I thought I had a good appreciation of their range of furriness. But the echidna I met in the Kimberley was bald. It was thoroughly spiny, but otherwise its skin was pink and naked. Looking closely, there was the sparsest spattering of little pig-bristles, each a few millimetres apart, and 5 millimetres ($^1/_5$ inch) long, but it had no fur.*

Tropical Australia has three indistinct seasons: the dry season, the wet season and in between, the 'build-up'. The build-up is the hardest to bear, as the temperature soars well above 40 °C (104 °F) and the humidity nears 100 per cent as the wet season approaches, but refreshing rain hardly ever comes. When we do fieldwork at remote camps during the build-up, we get up at around 4 a.m. to check our traps before the sun rises, and then lie sweating all day in a cave or any other natural shelter we can find, only re-emerging to reset the traps at around 4 p.m. On some trips, we have been in these boiling conditions completely surrounded by water, but unable to jump in and

---

*There could be a couple of reasons why naked echidnas are not well represented in European museums: firstly, that far fewer collecting expeditions have visited this region than other parts of Australia, and also because bare skin looks terrible when tanned and dried in museum specimens.

cool off as there were saltwater crocodiles everywhere (the sites from that trip with the echidna were too far inland for salties).

Echidnas cannot sweat or pant,[24] so it's fair to assume that the naked echidna's lack of hair is an adaptation to dealing with these unforgiving conditions. It must be serious for them to evolve this baldness, as thick hair is valuable protection from the biting jaws of all the ants and termites whose nests echidnas are constantly breaking into. I can say from personal experience that being covered with biting ants and dealing with the extremes of tropical heat and humidity at the same time is hard going, and if I had to choose to ease the discomfort of only one of them, I think I would opt to cool off too.

My surprise at encountering this unfamiliarly bald form of echidna, however, was nothing compared to the responses of eighteenth-century naturalists to the first monotremes to reach British shores.

4

The Mystery of the Egg-laying Mammals

Upon the first arrival of specimens in Europe, people found the platypus to be literally beyond belief. Initially, they were considered a possible hoax – a chimera of animal bits sewn together by Chinese con artists. 'Artist' is the operative word here, as European collectors had become wary of the exquisitely executed taxidermic fakes (such as 'mermaids') coming out of Asia at the time, which were sold at ports to sailors as they passed through. George Shaw, the man who provided the first scientific description of a platypus in 1799, was sceptical of the historic specimen he had before him ('Of all the Mammalia yet known it seems the most extraordinary'[1]). He even attempted to uncover possible fraud by subjecting the specimen to treatments that would dissolve any stitches or adhesives used to stick different animal body parts together. This notion presumably explains why, when visiting the museum storerooms that house some of the oldest specimens today, I have found the occasional detached bill with scrappy bits of dried flesh, skin, fur and unevenly snapped bones poking out. I suspect that they were ripped apart from the carcass – rather than carefully dissected, as one would expect in a more anatomical study – in an attempt to prove that it was only creative stitching that had attached what looked like a duck's bill to what seemed to be an otter's body.

Shaw was inspired to apply this quote from eighteenth-century French naturalist Georges-Louis Leclerc, Comte de Buffon, to the platypus: 'Whatever was possible for Nature to produce, has actually been produced.'[2] And when Darwin was in Australia, he speculated on the mammal's origins:

> In the dusk of the evening I . . . had the good fortune to see several of the famous Platypus . . . A little time before this I had been lying on a sunny bank, and was reflecting on the strange character of the animals of this country as compared with the rest of the world. An unbeliever in every thing beyond his own reason might exclaim, 'Two distinct Creators must have been at work'.[3]

*

Writers are constantly questioning their own originality. I was halfway through writing this book when I came across a quote from Harry Burrell that made me wince with recognition. He said, 'Every writer upon the platypus begins with an expression of wonder. Never was there such a disconcerting animal!'[4] I had used the word 'wonderful' in literally the first paragraph. But in his assertion, Burrell was also highlighting the zoological problems that the discovery of the platypus presented to European scientists. Platypuses broke the taxonomic rules about what features an individual species was meant to possess. There was much debate about where platypuses fitted on the tree of life. Were they furry, milk-producing reptiles? Amphibious mammals with reptilian bones? Missing links? Half-formed almost-mammals? Something that was neither reptile nor mammal?

Just as he had previously done with echidnas, in 1799 Shaw initially allied platypuses with South American anteaters. A year later, the German physician and naturalist Johann

Friedrich Blumenbach presented another theory – he had his own personal system of taxonomy for mammals, and placed the platypus within his group Palmata, which were 'the mammals with swimming feet'.[5] Specifically, Blumenbach thought it was 'an intermediate between the otter and the walruss [*sic*]'.[6]

Eventually, new taxonomic groups had to be created to accommodate them. In fact, several new taxonomic groups were proposed for what we now call monotremes. Every naturalist worth their salt had a theory, and each of them wanted to be the one to name the intriguing new group. Harry Burrell set the situation out as well as anyone could:

> Lamarck (1809) created a new class, Prototheria, for platypus and echidna, deciding that they were not mammals, for they had no mammary glands and were probably oviparous [egg-laying]; they were not birds, for their lungs differed, and they had no wings; and they were not reptiles, for they possessed a four-chambered heart. Illiger (1811) placed them in a division Reptantia, intermediate between reptiles and mammals. Blainville (1812), on the other hand, was convinced that they were mammals, though belonging to a separate order, Ornithodelphia. He was the first to indicate their many close points of agreement with the marsupials, and gave a long list of mammalian resemblances.[7]

None of them got it quite right – at this point no taxonomist knew enough about the animals' biology. Lamarck's choice of word is interesting: Prototheria means 'primitive or original beast', illustrating that notion that they were somehow lesser. Until relatively recently, the idea that monotremes and marsupials were closely related was a popular one (in line with Blainville's beliefs, above). There are some biological reasons behind this theory – for example, their babies start off life tiny, and they both have two bony struts projecting forwards from their pelvises – but it's also convenient for upholding the unscientific colonial philosophy that placental mammals like us are a cut above the beasts from down under. However, we now know that marsupials and placentals are each other's closest relatives, and they are both equally related to platypuses and echidnas. The characteristics which marsupials and monotremes share were present in the first mammals, so they aren't helpful in discerning relationships between modern mammals.

We shouldn't be hard on the men who made these conclusions more than 200 years ago. I have extreme respect for those pioneers working with a tiny handful of skins and carcasses in various stages of preservation sent from the other side of the world.* Not only that, but platypus biology varies dramatically depending on the time of year. No wonder Lamarck assumed they had no mammary glands. No one had ever seen an animal with mammary glands but no nipples, and unless they are lactating, the glands are minuscule.

Shaw's 1799 description officially introduced the platypus to the wider world. We don't know for certain the history of the specimen on which he based his account. The dates don't add up for it to have been the skin that Governor John Hunter acquired in November 1797 and sent to Newcastle (arriving there in December 1799. See pages 9 and 10).

It seems that a specimen left Sydney in August 1797 aboard the *Britannia*, a convict-transporter that travelled to Canton (Guangzhou) in China after it left Australia.[8] There, one of the *Britannia*'s officers apparently sold a platypus to an officer of the *Ceres* – one of the East India Company's ships, which arrived in London in October 1798. This passed to the physician William Buchan, of Store Street in London (a stone's throw from my former office in the Grant Museum of Zoology, but more importantly just around the corner from the British

---

* Although it should be noted that elements of Blumenbach's work on humans were later used in support of scientific racism.

Museum, where Shaw worked). Perhaps, *perhaps*, this is where Shaw procured his specimen.*

In any case, the first full anatomical description could not be given until complete preserved specimens were at hand, rather than just skins. Again, it was Sir Joseph Banks who was pulling the strings as the sole purveyor of platypus material in Europe. A year after that first skin, Banks received two whole platypuses – a male and a female – preserved in alcohol. At least one of these was sent to him by Governor Philip Gidley King, who had recently arrived in New South Wales to take over the role after Hunter returned to England.

Banks gave distinguished surgeon and Fellow of the Royal Society, Everard Home the honour of dissecting these specimens in order to describe their biology in glorious detail. Home proved worthy of the privilege. It truly is astonishing what he was able to observe. His dissection skills and his powers of deduction were outstanding. (He was later elevated to become Sir Everard Home, which I like to think was in some way recognition for his platypus work.†)

For example, Home discerned from his dissection that – despite the specimens' brains having decomposed to a point where he couldn't usefully describe their structure – the nerves which supplied the platypuses' faces were very large. From this he correctly deduced that the bill was a major sensory organ, and 'that it answers the purpose of a hand, and is

---

* The only note Shaw makes is that it 'is at present in the possession of Mr. Dobson, so much distinguished by his exquisite manner of preparing specimens' (Shaw, 1799). But we don't know how Dobson fits into the chain of transactions above. Adding further mystery, a century later the British Museum's Oldfield Thomas wrote, 'I have lying in front of me the original type skin of Shaw's *Platypus anatinus* but I am afraid I can give you practically no details of its history more than is printed in my Catalogue of Mammalia' (Burrell, 1927). That Catalogue indicates it was presented by 'R. Latham Esq.' (Thomas, 1888) – which suggests that either this is not the true specimen, or Shaw did not acquire it for the museum when he described it.

† Although he also served as Sergeant Surgeon to King George III. You can make your own mind up which was the greater factor.

capable of nice discrimination in its feeling'.[9] Incidentally, Blumenbach had reached a similar conclusion from the skull and dried remnants of tissues inside the skin he studied in 1800 (also sent from Hunter to Banks),[10] which is perhaps more impressive – certainly suggesting that Banks' choices for who should work on these specimens were well judged.

At this early stage in platypus science, the big question of where to place the species taxonomically was already at play. The mode of reproduction was a major deciding factor and, with a female specimen at his disposal, Home sought to share the evidence: 'There is no appearance, that could be detected, of nipples; although the skin on the belly of the female was examined with the utmost accuracy for that purpose.' And on the reproductive system itself: 'This [overall] structure of the female organs is unlike any thing hitherto met with in quadrupeds'.[11] While Blumenbach had somehow concluded that platypuses laid eggs (which, again, is remarkable considering he was only working with what he could see from dried skins), Home came to a different conclusion. He thought that they most closely resembled species of lizards and sharks whose females produce eggs that then hatch inside the mother's body before she gives birth to live young. This is known as ovoviviparity, while egg-laying is oviparity and live-birthing is viviparity.*

And thus, the great controversy was underway. How did the platypus and echidnas reproduce? Could an animal be a mammal if it didn't give birth to live young? They had fur, which only mammals had. But they lacked nipples, which no mammal was so far known to lack. It became one of the most pressing zoological questions of the era.

*

---

*The term 'live-birthing' is problematic, as obviously eggs are alive too. But we don't have sensible alternatives.

By this point, of course, Aboriginal Australians had known the platypus for tens of thousands of years, which any statements about the 'discovery' of the animal, or of any Australian species, fail to recognise. The platypus has different names in different Indigenous languages, including *malungan* and *biladhurang*, and perhaps predictably for a species whose range extends over a couple of thousand kilometres, there are different reports regarding whether platypuses were traditionally hunted as human food. Nineteenth-century writers in Queensland suggested Indigenous people there did not like platypus flesh,[12] but others claimed that people from near modern-day Canberra did eat them;[13] there are accounts from Tasmania of people habitually digging them out of their burrows,[14] presumably for food; and the platypus sent in the barrel by Hunter to Newcastle was caught by a Darug man apparently in the course of his everyday hunting habits. In any case, they certainly feature in Aboriginal cultures and storytelling.

David Unaipon, a Ngarrindjeri writer and inventor (who currently features on the Australian $50 note), retold a platypus legend from the Berrwerina people of the Darling River, in his 1920s collection of traditional Aboriginal stories. It begins with the leaders of the mammal, bird and reptile tribes calling a meeting in the Blue Mountains. They had met to discuss how to deal with the fact that all the creatures were reproducing too quickly and were crowding each other out.[15] It was remarked that the platypuses were so numerous that the other animals were constantly tripping over them. While the leaders were discussing solutions for the problem, the selfish Frilled Lizard decided to take matters into his own hands and called upon Lightning, Thunder, Rain, Hail and Wind, asking them to bring a flood that could destroy the platypuses.

The flood came and demolished everything in its path. The birds flew away to safety, and the mammals and reptiles ran to seek refuge on the mountaintops. Once the waters subsided, the mammals went in search of the poor platypuses – and mourned their loss when all they found were dead bodies. Hearing their cries, the Frilled Lizard told the Kangaroo that they had no reason to be sad, as the platypus was not related to the mammals, birds or reptiles.

Once the birds returned, three years later, they too hoped to rediscover the platypus. The Cormorant excitedly reported finding foot- and beak-prints in the mud by a pool, halfway between the Blue Mountains and the coast.

On hearing the news, conscious of what the Frilled Lizard had said, the Kangaroo – the leader of the mammals – arranged a great gathering and asked the birds, reptiles and mammals to look back among their ancestors to find out how closely related they might be to the platypus, in the hope that this would make the other animals want to help them.

The birds were the first to uncover a relationship – the Black Duck realised that platypuses have the beak of a bird, and that one of their relatives – the wigeon – has a beak with comblike 'teeth', just like the platypus. With this, the birds swore to protect the platypus. Next, the Pelican noted that he must be related to platypuses, as they lay eggs – the Kookaburra pointed out that they were more like reptile eggs, but the others said that it didn't matter. The Emu thought it was a shame that platypuses did not have feathers, as they would make nice birds.

When the reptiles studied their family tree, they too noticed the similarities with their eggs; and the mammals – with their fur – decided that they were definitely relatives of the platypus.

Since the platypuses were a wise and ancient race, the animals decided that they should be asked what they thought.

And so, the Platypus elder was asked to join the other animals at their meeting. When he arrived, he told the crowd that long ago, he was related to the reptile family, and then later he became related to the mammal family, and after that the bird family. Although he had lost the feathers of his ancestors, his wife lays eggs and he has a beak, so he could never forget he was related to the birds; and because he has fur, he could never forget he was related to the mammals. There were jealous arguments between the other tribes over which of them the platypus should join, but in the end, Platypus just wished to be a platypus – not a bird, a mammal or a reptile.[16]

A similar story from the Central Coast of New South Wales, retold by storytelling custodian Pauline McLeod, involves the mammals, birds and water creatures arguing over which of these three groups are the most important. Each of them recognises their affinities with the Platypus, and each harasses him to join their group. After much deliberation (and consultation with the Echidna), Platypus decides not to join any of their groups, as he is special in his own way, saying, 'I don't know why the ancestors have made us all different, but we must learn to accept all these differences and live with each other.'[17]

Elsewhere, in northern New South Wales, a platypus creation story retold by Aboriginal educator June Barker appears to serve as a warning to children who disobey their elders and wander too far from home, or marry those who are not picked by their parents. A young duck, Gaygar, swims too far away from her lake and is kidnapped and raped by Bigoon, the spear-carrying water rat. She manages to escape and return to the lake, but the other ducks' celebrations turn sour when it comes to egg-hatching time. Her babies have ducks' bills, but four webbed feet instead of two, dark brown fur instead of feathers, and on their hind feet are spurs like Bigoon's sharp spear. To stop the other ducks killing her babies, Gaygar is banished from the lake and flees. She eventually dies of a broken heart, but her young – the first platypuses – thrive in their new cold-river home.[18]

These Indigenous Dreaming stories (the Dreaming is the period of creation before time as we know it existed), like European perceptions, clearly share the view of the platypus as some kind of hybrid, chimera or in-between species, combining features of different groups – and are unified by the notion that they are duck-billed. They also all demonstrate that at the time of telling, platypuses were recognised as laying eggs.

*

Given how eagerly an answer to the mystery of platypus reproduction was sought in Europe, one might imagine that information from the people who knew the land and its fauna best would be highly valued.

In their exceptional 2019 book on Indigenous Australians' contributions to early zoology, *Australia's First Naturalists*, Penny Olsen and Lynette Russell suggest that the English naturalist George Caley was the first to recognise that value. His is the earliest account known to demonstrate that Aboriginal people told the colonists that platypuses and echidnas lay eggs.[19]

Caley arrived in New South Wales in 1800 in the employ of Joseph Banks. He had been sent there specifically to gather knowledge and specimens of the fledgling colony's natural history. This arrangement would last for eight years before Banks approved Caley's return home. Caley had made the journey from England aboard the HMS *Speedy* with Governor King. Caley even stayed with King in Government House when they first arrived. Both men had been mentored by Banks and would write to him regularly with updates throughout their time there.

Perhaps reflective of their different roles in the new settlements, Caley and King had very different approaches to relationships with the Indigenous population. King and the people working under him used extreme violence to attempt to suppress Indigenous resistance to white settlement.* Olsen and Russell describe how Pemulwuy, the Eora resistance leader

---

*I always find 'settlement' to be a thoroughly unsuitable word, as if it were as gentle as settling snow.

(Eora was the name given to the people whose country includes the areas that the British first settled, around what is now Sydney), had been shot dead in June 1802, and King had sent his head as a trophy to Banks. He wrote, 'understanding that the possession of a New Hollander's head is among the desiderata, I have put it in spirits and forwarded it by the *Speedy*'.[20] This suggests that King and Caley had effectively been sent to Australia with a shopping list of things that would be of interest to Banks, and that list included an Aboriginal person's head.

Caley didn't want any of this unrest and made clear his intention to establish good terms in order to be able to exchange information about plants and animals in return for food and tools. Within two years he had learned enough language to communicate with a number of the local Eora groups and, to an extent, mixed freely with them. He protested that conflicts with 'the natives' – which he considered to mostly have been instigated by the colonisers – were hampering his ability to build the relationships so vital to collecting facts and material to send back to Banks. Caley clearly appreciated the value of the natural historical insights of the Darug and Eora people, and regularly mentioned by name such individuals who had contributed to his understanding – a mark of respect that was often not practised by colonial naturalists the world over. He also told Banks that he 'could single out several that surpass numbers of Englishmen in mental qualifications'.[21]

Elsewhere on that list of '*desiderata*' was information on the platypus and echidna. Caley wrote to Banks in 1803:

I have made strict enquiry how they generate; I asked several natives but could not receive any satisfactory answer, but still continued asking the different natives that I chanced to see. At length I learned from one man what I think may be relied on, who told me that 'they went a long way underground and layed eggs'. The Porcupine Ant Eater [echidna] is entirely

unknown to me and by what I can learn from the natives
it is scarce.[22]

In the following year he received another report from an
Indigenous source that backed up the claim that platypuses
laid their eggs underground. Likewise, Burrell quotes a letter
from 1822 from the surgeon Patrick Hill that demonstrates
that Indigenous knowledge was being gathered that would
help to solve the platypus mystery – with some added commen-
tary reflecting the common problem of details being lost in
translation:

> 'Cookoogong a native, chief of the Boorah-Boorah* tribe,
> says, that it is a fact well known to them, that this animal
> lays two eggs, about the size, shape, and colour of those of
> a hen; . . .' As the egg of the platypus measures only three-
> quarters of an inch in length, Cookoogong's estimate of the
> size is rather wide of the mark; but this may have been due
> to misunderstanding on the part of Hill.[23]

In the 1830s, platypus pioneer George Bennett made several
enquiries for information on their reproduction to Indigenous
people in the areas in which he was undertaking zoological
collecting in New South Wales, but was unable to overcome
the language barriers in order to draw satisfactory conclusions.

From these and other accounts, it's clear that at least some
Europeans with an interest in natural history were asking
Aboriginal people for information about Australian wildlife.
However, in unfathomable displays of the classic colonial
superiority complex, many European scientists dismissed out
of hand the knowledge that was gathered. Many continued
to argue against the idea that monotremes lay eggs.

---

* Presumably a reference to the Barapa Barapa people, whose country covers areas
of southern New South Wales and northern Victoria.

This failure to incorporate Indigenous knowledge into the Western scientific understanding of Australia's natural history went beyond the platypus and echidna. Olsen and Russell make the point that while several species only became known as a result of specimens provided by Aboriginal collectors, unrepeatable opportunities to gather information were squandered, and then it was too late. It didn't take long for animal populations around new settlements to crash, and for the original human inhabitants to disappear from the areas – and those that remained had significantly different relationships with the environment due to the presence of the Europeans. Some species – particularly those endemic to islands off Australia's southeast coast – were quickly lost entirely.[24]

*

The battle over whether monotremes lay eggs was deeply controversial and took nearly a century to settle. The main opponent of the egg-laying theory was the brilliant but dastardly Richard Owen, who believed that they produced eggs internally, but that they hatched inside the mother's body and then she gave birth to live young (i.e. they were ovoviviparous). Owen was one of Europe's leading anatomists, whose contributions were many and major. He established what we now call the Natural History Museum, coined the term 'dinosaur' and described many iconic extinct species from molluscs to moas (including several of the most famous giant fossil marsupials). He was also prone to claiming other people's discoveries as his own, stealing other scientists' deliveries of specimens as they arrived at British ports (allegedly) and strategically destroying his rivals' careers in order to decrease competition for scientific recognition. He was, by all accounts, Machiavellian, jealous, egotistical, malicious and spiteful.*

---

* A *bad egg*, if you will.

But don't take my word for it. Charles Darwin wrote angrily to his friend Joseph Dalton Hooker, director of the Royal Botanic Gardens at Kew (who, incidentally, Owen had been bad-mouthing, apparently as a ploy to seize control of Kew's collection for himself by bringing it under the umbrella of the British Museum), declaring, 'As far as I understand things, nothing equals Owen's conduct. I used to be ashamed of hating him so much, but now I will carefully cherish my hatred & contempt to the last day of my life.'[25] Earlier, Darwin had also let slip his dislike for Owen in this snarky comment to Asa Gray, an American botanist at Harvard University: 'No one fact tells so strongly against Owen, considering his former position at [the Royal] Coll.[ege] of Surgeons, as that he has never reared one pupil or follower.'[26]

Despite these opinions, Owen was well connected and powerful in the major European scientific institutions of the time. No one could deny his anatomical mastery. Owen described in painstaking detail the reproductive anatomy of five preserved female platypus specimens he had acquired, with the intention of confirming that they had milk-producing mammary glands – and therefore were certainly mammals – but that they didn't lay eggs. A key target of his study was to prove that a set of glands under the skin of the female's abdomen – which were not found in males – truly produced milk.[27]

The specimens Owen dissected had been caught at different times of year and so gave him the opportunity to compare them and make deductions about how female platypuses might change across the course of their reproductive cycle. He found that the size of the glands differed widely and were largest in the specimen whose ovary appeared to have recently released an egg, as one would expect if their function was to produce milk for a newborn.

There's no doubt that Owen was dedicated to his pursuit of proving platypuses' milkiness. Selecting the specimen that he suspected of nursing young at the time of its death, he squeezed the glands, and studied the tiny holes where they opened onto the skin. '[T]here escaped from these orifices minute drops of a yellowish oil, which afforded neither perceptible taste nor smell, except such as was derived from the preserving liquor.'[28] That's right. Richard Owen tasted the milk from a platypus that had been dead in a barrel of alcohol for several months, if not years. I am now left curious as to how many other people have drunk platypus milk over the course of history.

In any case, the rest of his findings satisfied him that these glands were the equivalent of the mammary glands found in other mammals, and did indeed produce milk. He then turned to the question of how the babies could feed without nipples. He found what he thought were muscles around the glands, which could contract to squeeze them against the ribs, thereby squirting the milk out onto the skin, so the babies had no need to suck.

On the topic of milk, Richard Owen was, we now know, correct. On the other hand, he did not reach an accurate answer on the topic of their mode of reproduction from his work on these five specimens. He finished with the same opinion he started with: that platypuses could not lay eggs. To me, there is little in his anatomical descriptions of the reproductive system that would have helped draw a conclusion one way or the other. From what he saw, he didn't think the walls of the canals could secrete a shell. But there appears to be little more than that to have guided his conclusion. This is of course not surprising, as the conclusion was wrong.

Just as Owen's paper was going to print, the Zoological Society of London – at a meeting that Richard Owen was himself chairing – heard a very brief report from a Dr Hume Weatherhead, which answered the questions they had all been struggling with far more simply. Weatherhead had received a letter from Lieutenant Lauderdale Maule in New South Wales. It added serious weight to the evidence for milk production and egg-laying in platypuses. Owen tacked it onto his mammary glands paper as an appendix. The report stated:

> 'Several . . . [platypus] nests were with considerable labour and difficulty discovered. No eggs were found in a perfect state, but pieces of substance resembling egg-shell were picked out of the debris of the nest. In the insides of several female Platypi which were shot, eggs were found of the size of a large musket-ball and downwards, imperfectly formed however, i.e. without the hard outer shell, which prevented their preservation.'
>
> In another part of his letter Mr. Maule states, that . . . on skinning [a female] while yet warm, it was observed milk oozed through the fur on the stomach, although no teats were visible on the most minute inspection; but on proceeding with the operation two teats or canals were discovered, both of which contained milk.[29]

There are three key discoveries in this short report: eggshells, internal eggs without a hard shell and oozing milk. The eggshells contradict Owen's views, the meaning of the internal eggs is ambiguous and the milk supports Owen's position. What Owen does is dismiss the part of the letter that disagrees with his own theory, strategically interpret the ambiguous bit and accept the part that helps him. Owen already thought that platypuses produced eggs that hatched internally, so he said the eggshell was probably poo. And on scant evidence he decided that the 'eggs' Maule described from within the females' tracts sounded very much like those in the ovoviviparous reptiles that hatch inside the mother. However, Owen thought that the story about the milk was obviously true.

Two years after his initial project, Owen got his hands on more specimens – largely thanks to George Bennett in New South Wales – allowing him to dissect and describe a lot of eggs from within the bodies of female platypuses. They were a range of sizes, reflecting different stages of development. He noted that in none of them – even the large ones – could he find any trace of an embryo.[30] This should have rung alarm bells for any unbiased scientist considering the question of whether the species was ovoviviparous: if the egg were going to complete its initial development within the mother and then hatch before the baby was born 'alive', shouldn't the largest eggs within the mother have an actual foetus inside them? Instead, he just found various fluids, tiny granules and thin membranes.

There were two possible explanations for these findings: first, that the eggs were laid as eggs, and completed their development outside of the mother's body, through incubation, before hatching (we now know that this is correct); or second, that they had yet to find specimens with eggs that had been developing in the mother long enough for the embryo to grow to a sufficient size for it to 'hatch' internally before its birth. Despite the impressive number of specimens so far discovered with internal eggs, Owen made the stretch to land on the second option.[31] He stuck to his guns and continued to argue that platypuses did not lay eggs. He went on to instruct Bennett to shoot a female platypus every week during the breeding season until they had a series of specimens covering every stage of development, but over time Bennett became anxious that wholesale slaughter of monotremes could lead to their extinction.[32]

In 1864, Owen received a letter from Dr Jonathan Nicholson in Victoria, outlining what could so nearly have been the end of the matter. A female platypus had been caught on the Goulburn River and secured in a box overnight. In the morning, there were two eggs, 'about the size of a crow's egg, and were white, soft, and compressible, being without shell or anything approaching to a calcareous covering'.[33] Sadly, when Nicholson checked on them a day or two later, they had been thrown away.

Owen was understandably frustrated by this lost opportunity, but in his account of the matter he does not accept it as true and appears to suggest the eggs were 'abortions caused by fear'.[34]

Brilliant anatomist as he was (aside from the platypus egg dispute), Owen was on the wrong side of some of the biggest biological debates in history. The chief of these was natural selection, and he is remembered as being the foremost critic of Darwin and Alfred Russel Wallace's revolutionary theory. When it came to the possibility of platypus eggs, I think the reason that he denied their existence is that his staunch Christian conservative views made him uncomfortable with an animal that was clearly a mammal – a member of our own noble class – doing something so 'primitive' as laying eggs. It dragged us mammals down into the dirt with the belly-dragging reptiles, which Owen considered a challenge to our moral superiority. On top of that, those with socially conservative views, like Owen, were eager to demonstrate that animal groups were unchanging and confined – they were established by God and one could not cross into another, just like people from one social class could not cross into another.[35] The political desire to establish the absoluteness of such hierarchies was still at the forefront of people's minds following the French Revolution, and the upper echelons of society saw danger in scientific evidence for the idea that things could change. If platypuses demonstrated transitional characteristics between reptiles and mammals, that absolutist argument would be weakened. If things could change in nature, they could change in human society, too. And that didn't suit Owen and his elite patrons at all.

Owen was wrong, and when it finally came to it, the evidence needed to prove to sceptics in Europe that platypuses truly did lay eggs would not have been found without Indigenous knowledge.

5

Cracked

It wasn't until 1884 – some 85 years after the first specimen arrived in London – that the bright young Scottish embryologist William Hay Caldwell, working with Aboriginal collectors, eventually settled the egg question to the satisfaction of European scientists.

Caldwell had been awarded a first-class degree in natural sciences from the University of Cambridge in 1881 and began his post-graduate studies in embryology under the mentorship of his compatriot Francis Maitland Balfour. At a young age, Balfour had become a shining light of British biology with a worldwide scientific reputation. However, shortly after being promoted to professor, he contracted typhoid. For the final stage of his recovery, he sought the fresh air of the Alps, and attempted to become the first to climb the Aiguille Blanche de Peuterey, an unscaled peak of Mont Blanc.* On or around 19 July 1882, he fell and died, aged just 30.[1]

William Caldwell was then the university demonstrator in comparative anatomy – teaching advanced lectures on invertebrate biology to third-year students only two years younger than him. 'Lecturing was not his *forté*,' one of his students,

---

* I've had typhoid, and trying to establish a new mountain-climbing route was not something I considered as part of my own recuperation process. Balfour was clearly a determined man.

George Parker Bidder, would write many years later, but, 'he had a charming deprecatory smile'.[2]

He was excelling in the lab, showing not only his biological prowess but also a penchant for engineering. At age 22 he invented the Caldwell Automatic Microtome as a means for cutting extraordinarily thin slices of embryos (or anything small) for use in microscopy, doing so in a way that each slice sticks to the edge of the slice before, so it creates a ribbon of sections in the correct order as you travel through your object of study, all facing the same way. The device was motor driven, enabling 200 slices to be cut each minute.[3]

Not long before he died, Balfour had suggested to Caldwell that he should find the opportunity to take two or three years away from Cambridge to study the embryological development of the platypus, echidna and lungfish in Australia. Much like the monotremes, lungfish have long been a focus for biologists as they represent a significant evolutionary story: they are fish with limb bones like land animals, and the ability to breathe air. Unravelling the secrets of how these three animals grew would represent a major scientific development. As it happened, it was Balfour's climbing accident that enabled Caldwell to make the journey. He was awarded the first 'Balfour Studentship', set up by the university to commemorate Balfour (my department still offers it to support a new PhD student each year), which seed-funded a major expedition.

Caldwell reached Australia in September 1883. Demonstrating the perceived importance of the platypus question, he had been afforded every opportunity. Aside from money from Cambridge, he had funds from the UK government and the Royal Society. The Colonial Office had given him letters of introduction to the governors of New South Wales, Victoria and Queensland, and the British Admiralty did the same to the naval commander-in-chief of all Australian waters. Shortly after he arrived, the Australian government provided him with laboratories and chambers in Sydney. All of this at the age of

24. He must have felt like he owned the place (like pretty much every colonial envoy, I suppose).

Straight away, Caldwell set out to gather information on platypuses in New South Wales, but he struggled to find enough, because 'the skin-hunter had been before him',[4] referring to the impact of the fur industry on platypus numbers. And besides, he soon realised that it was not the right time of year for platypus eggs and embryos, so he collected koalas and other marsupials and prepared embryological specimens from them. At the time, the early stages of marsupial development were not well-known, and although Caldwell is most famous for his monotreme discoveries, between late 1883 and early 1884 he gathered the embryological stages of marsupials from impregnation up to birth, which no one had done before.

In mid-April 1884 he moved north to the Burnett River in southeastern Queensland. Caldwell was not one of the colonial scientists who dismissed the value of Indigenous knowledge (although it's worth noting that he doesn't credit the contributions of any individual Aboriginal collectors by name). He wrote, 'The Burnett district presented the further advantage of still possessing a considerable number of black natives. I afterwards found that without the services of these people I should have had little chance of success'. Olsen and Russell identify the Waka Waka as the Aboriginal group whose country Caldwell was working in.[5]

At camp, he had a massive team of over fifty Waka Waka people working for him in search of breeding animals. Things went well: he wrote that the women collected lungfish, and the men focused on 'their favourite food', echidnas (he struggled to encourage either to search for platypuses). Between April and December 1884, he built up a complete series of specimens demonstrating the stages of embryological development of lungfish, as well as many of the stages for monotremes. This is where monotreme history is made.

Caldwell elaborates on his findings, but only slightly. In

typical no-nonsense fashion, this is how he ended the great mystery of the platypus, in his own words:

> Towards the end of July [1884] the blacks began to collect Echidna, and very soon I had segmenting ova from the uterus [i.e. eggs that were dividing to become embryos]. In the second week of August I had similar stages in [the platypus], but it was not until the third week that I got the laid eggs from the pouch of Echidna. In the following week (August 24) I shot a [platypus] whose first egg had been laid; her second egg was in a partially dilated os uteri [the mouth of the uterus]. This egg, of similar appearance to, though slightly larger than, that of Echidna, was at a stage equal to a 36-hour chick.[6]

This is a widely quoted passage in monotreme lore. Caldwell's confirmation of egg-laying in platypuses – through the smoking gun of one egg in the nest and another about to be laid – is commonly said to be the moment the monotreme battle ended. With every retelling of this tale, I can't find a single one that highlights the fact that Caldwell did not actually answer the question on 24 August 1884 with that platypus – but instead he did it the week before, beginning Monday 18 August. That's when he states, 'I got the laid eggs from the pouch of *Echidna*'. Caldwell's first account of monotreme eggs was in an echidna, not a platypus. I wonder why this has apparently gone unnoticed. Is it because platypuses are more famous? Or, did this historical mix-up happen because Caldwell never made a big deal of the breakthrough himself – offering no flourishing fanfare of the momentousness of this discovery – allowing others' hyped-up versions of it to overtake the original? He never describes it as a eureka moment, or as a threshold in proving that some mammals do, in fact, lay eggs.

To him, it seems that the greatest significance of that platypus was that it allowed him to create a benchmark in mapping the stages of their embryonic development. He 'had discovered by a lucky chance . . . one which had laid one egg, and had the other on its way out through the passage'.[7] This meant that those eggs would tell him what a platypus embryo looks like at the point it leaves the mother's body. Seeing as all descriptions of infants and embryos are in relation either to their age since fertilisation (when they are inside the mother), or age since birth or laying (when they are outside), the appearance of the embryo at 'day zero' would unlock the means to describe the embryos of all other platypus eggs in comparison to that point in time.

Caldwell appears most proud that his discovery enabled him to announce that the platypus embryo was at a similar stage of development when it was laid as a hen's chick is thirty-six hours after being laid – i.e. that platypuses are more developed at the point of laying than chickens are.

With that said, he did at least realise that the question the world was waiting for was the big one: eggs or no eggs? And so, on 29 August, with perhaps the most famous telegram in the history of zoology, Caldwell made an announcement in words few laypeople could understand: 'Monotremes oviparous, ovum meroblastic'. In an 1884 parallel to scientists today who post their discoveries live from the field on Twitter, that was the message Caldwell sent to a meeting of the British Association for the Advancement of Science which was taking place in Montreal.*

The first two words confirm that platypuses and echidnas lay eggs, and the second two mean that their eggs have large yolks, upon which the embryos develop in the same way as birds and reptiles (they are meroblastic). He couldn't have used any fewer characters to tell this story.

---

* If you're wondering how he could telegram Canada from the bush, he couldn't: he telegrammed a neighbouring station, where the mailman picked it up. The postal service then delivered it to his friend, Professor Liversidge, in Sydney, who arranged for it to be telegrammed to the meeting in Montreal.

The telegram caused quite the stir at the meeting when it was read on 2 September:

> The President stated that he had a most important announcement to make. He had just received a cablegram from Sydney, from Professor Liversidge, announcing that Mr. Caldwell . . . had discovered that the Monotremes were oviparous. He did not consider that a more important telegram in a scientific sense had ever passed through the submarine cables before.[8]

In a genuinely remarkable coincidence, on the very same day a similar announcement was being made in Adelaide. Wilhelm Haacke, a German zoologist and director of the South Australian Museum, was sharing big news with the Royal Society of South Australia: 'I found an egg in the mammary pouch of the female, and was thus enabled to prove that *Echidna* is really an oviparous mammal'.[9] And he had brought the egg with him to exhibit.

His discovery had been made a week earlier on 25 August, with an echidna that had been brought to Adelaide from Kangaroo Island. That was the day after Caldwell had found his famous platypus, and a week after Caldwell's similarly egg-bearing echidna. There is no question that Caldwell was first – but it is astonishing that after eighty-five years of Europeans deliberating whether monotremes lay eggs, two people would reach the answer at essentially the same moment. While Caldwell has priority, Haacke did at least give the discovery the honour of spelling out its significance.

Any story about the history of platypus science pays homage to Caldwell, but few mention Haacke. Why is that? On the face of it, Caldwell *was* first, and to the victor go the spoils (although Caldwell never showed any signs of seeing it that way). Also, Caldwell's discovery went beyond the simple fact of egg-laying – Caldwell made extensive discoveries about monotreme embryology, which presumably elevates the scientific value of his work. By contrast, when Haacke found his egg, it was dead and rotting in the pouch (the mother was still alive, though), presumably on account of the journey it had been subjected to. When Haacke picked it up, it burst, so he couldn't even take measurements or perform much in the way of useful science, beyond stating the fact of its existence (he couldn't find any trace of an embryo). Caldwell never took the time to describe the pouch, while Haacke provides us with this nugget: 'we find in my specimen one deep pouch large enough to hold, although not wholly to conceal, a gentleman's watch'.[10]

But are there political reasons why Caldwell gets so much more historical limelight than Haacke, at least in the English-speaking world? Caldwell was British, and Haacke was not, and that is likely to influence certain versions of history. Not only that, but Haacke's discovery was Australian, as he belonged to a colonial institution, while Caldwell's was British, as an envoy of the great imperial monoliths of the University of Cambridge and the Royal Society.

Nevertheless, both gentlemen acknowledged the other's work, and it's clear that Caldwell himself didn't agree with all the fuss. Their actual mode of reproduction didn't interest him, just how their embryos developed. He wrote to Professor Liversidge shortly afterwards:

[Lungfish are] much more important. Platypus embryos are quite easy to get. I can't understand how they have not been got before. The fact tha[t] the monotremes are oviparous is the end of the research for many. They don't understand that it is the fact of the eggs having a lot of yelk [yolk] that promises to yield valuable information.[11]

Caldwell here is rather disrespectful to those Australia-based naturalists who had dedicated much of their lives to solving the question that he dismisses. The letter was read at the

Royal Society of New South Wales. How must George Bennett have felt? He had worked on this for over fifty years, and then Caldwell swoops in from Cambridge, employs enormous numbers of Indigenous workers to collect for him, and gets the proof in a matter of months after his arrival. (Bennett, for what it's worth, did show some level of dissatisfaction with the situation, having written to Owen remarking on Caldwell's sizeable workforce, while he himself only 'had two lazy aborigines'.[12])

Caldwell returned briefly to Sydney and visited the Royal Society of New South Wales to make his first official scientific report. It makes for painful reading. Although the account of the meeting doesn't say so in so many words, Caldwell's privilege appears to have gotten the better of him, and he comes across as arrogant and dismissive.

The report gives us an insight into Caldwell's matter-of-fact scientific mind. He told the society that he had come to explain the embryology of marsupials, monotremes and lungfish, and basically, that's all they were going to get out of him:

> What he would state with regard to his investigations were not theories but were facts, and were consequently not open to argument. Within the last few weeks he had received several letters from people denying that the platypus laid eggs, and they wanted him to argue about it. That was impossible. He stated a fact; it was possible to disbelieve it; but, being a fact, it could not be argued.[13]

It gets worse. Someone at the meeting asked him whether he had worked out how marsupial babies get from the vulva to become attached to the teat. The answer to this question is, I think, one of the most astonishing feats in the natural world: a baby that may be significantly smaller than a grain of rice, after a pregnancy of a handful of weeks, climbs hand-over-hand, unaided, into the pouch. It. Is. Amazing.

However, this is what happened at the meeting:

Mr. Caldwell . . . said that the exact mode in which the kangaroo or other marsupial put the young to the teat was not of so much importance as the other facts. . . . He had not personally observed how the embryo was actually moved into the pouch — he had not considered it of sufficient importance to waste any time about. He could conceive no difficulty in the lips or tongue of the mother kangaroo placing the young . . . upon the teat. The question did not appear to him to be a matter of any importance – it did not form part of his researches.[14]

Can you imagine going to the learned society of another country, being asked a question about that country's wildlife by one of that country's scientists, and then dismissing the question as unimportant?

This was not the first time Caldwell had been surprised at what other biologists were interested in, and I suspect it betrays an arrogance of the value of his biological field over others. Looking through his accounts of all that time he spent in the field, collecting some of the most enigmatic species on Earth, because of the discoveries he does make, it's easy to miss that he makes essentially no mention of their natural history. He doesn't describe much behaviour, never describes their burrows – despite regularly accessing them – nor anything about their venom or diet or feeding. I don't like to think of science working in silos, but this is an embryologist tackling a question of embryology, with no interest in its zoological interpretation. Indeed, the society notes of that day:

The interpretation of these facts he was not prepared to add, as he had come there with the simple intention of exhibiting a few specimens, and not with the intention of entering into any theoretical consideration derived from these facts.[15]

Caldwell was apparently oblivious to – or entirely uninterested in – the wider implications that the fact some mammals lay reptile-like eggs would have for our evolutionary understanding.

Caldwell told the Royal Society of New South Wales that there was still a lot of work to do. Having done so for lungfish, he wanted to collect and describe a complete developmental series of every stage of platypus and echidna embryos, and that would require an even larger team of Indigenous helpers than before. And so he headed back into the field.

The next year did not start well for him. He spent the first three months of 1885 collecting and preparing specimens of marsupial embryos, but these were all washed away when his buggy overturned at a river crossing. He was then out of action for two months with a fever he had picked up on the Burnett River, 'and it was not until June that I started to organise the large camp of blacks, to continue the delayed attack on the Monotremes.'[16] If the language of warfare was called upon, perhaps it is because he had essentially employed an army. Over 150 Waka Waka worked for him, and they slaughtered between 1,300 and 1,400 echidnas in the name of science, enabling Caldwell to put together the embryological series he so desired. Except for a handful of cases, it's normally assumed that scientific collecting happens on such a small scale as to have negligible effect on animal populations, but at least at a local level it's hard to imagine how the echidnas could have coped with this onslaught. There's also reason to suggest that most of them were wasted: 'A skilful black, when he was hungry, generally brought in one female Echidna, together with several males, every day. The former seemed to be much more difficult to find than the latter at this season.'[17] He never mentions what he does with the males (since only the females have eggs and embryos), suggesting they died for nothing at an astonishing rate.

Caldwell paid his collaborators in cash for their efforts,

which was certainly not always the case with other European collectors, but he ensured that he retained the upper hand in the transactions. 'The blacks were paid half-a-crown for every female, but the price of flour, tea, and sugar, which I sold to them, rose with the supply of *Echidna*. The half-crowns were, therefore, always just sufficient to buy food enough to keep the lazy blacks hungry.'[18] Gathering 1,300 echidnas in three months does not seem particularly lazy to me.

Presumably, since they were based in a remote camp, Caldwell offered the primary means by which the Waka Waka could actually spend their earnings, so he was able to control the immediate economy by devaluing the cash he was paying them in order to maintain a market for echidnas.[19] Echidnas effectively became the currency of trade. All he had to do to maintain a supply of specimens was ensure that the amount of tea, flour and sugar they could buy for half a crown was of greater value than eating the echidna itself.

With echidna battle won, in September 1885 Caldwell travelled south to the Mole River (so named for the duck-mole) but became ill again. He couldn't complete the collection of all the stages of platypus growth as a result, and so would have to rely on the echidna material to describe the later stages of monotreme development.

In December 1886 he finished his studies in Sydney and got married before returning to England with his new wife, Margaret Gilchrist. It wasn't long before Caldwell was in front of the Royal Society in London, in March 1887, giving his account of the findings that it had helped to fund. His work was excellent, describing in great detail the composition of the monotreme egg and its embryonic cells and membranes as it develops from the ovary, through the uterus and to laying. He provided the same for marsupials and showed that before birth, their embryos have a shell covering comparable with that of monotremes, but only 0.01 millimetres thick.

His groundbreaking presentation, representing what he

considered to be initial findings, was published with the title 'The embryology of monotremata and marsupialia.–Part I'.[20]

There would be no Part II. Caldwell returned to Cambridge with his specimens, where he worked on them for years without announcing his results. In 1893, he moved to Scotland, having inherited a paper mill, and he never published anything more in biology.

Nonetheless, through his efforts, Caldwell had more or less settled the matter of egg-laying in monotremes. That it took so long was a symptom of pre-existing assumptions of how a mammal should look and behave. The notion of primitiveness was therefore attached to the platypus at the very beginning of its European scientific history.

Science has often been a key colonial enterprise. During those early stages of the British colony, the science was exported back to the UK along with the specimens, rather than taking place in Australia itself. Scientific patrons in the UK, such as Joseph Banks and Richard Owen, sought to ensure that they pulled the strings from 'home', rather than enable the settlers and Indigenous experts to develop their own scientific industry. Although there were institutions such as the Royal Society of New South Wales, the system was weighted to ensure the hierarchy was upheld, whether that was by keeping short leashes on Australia-based collectors or by British naturalists being temporarily deployed in Australia to seek certain answers or explore certain regions before returning home to write up their discoveries in the British publishing industry. Australia could provide the data (the physical work), but Britain would develop the theories (the intellectual work).

At the time, this may have looked like scientific apathy from the colonists. Indeed, in 1927 Burrell complains that he was unable to find any local accounts of platypuses from the early nineteenth century: 'We owe nearly all our early knowledge to English officials and visiting French naturalists. While scientific Europe thirsted for enlightenment, the colonist went blandly on with his pioneering.'[21] However, in her fantastic account of colonial science history, Ann Moyal argues that this is more to do with the fact that the likes of Joseph Banks were maintaining somewhat of a monopoly on Australian zoology.[22] On top of that, discoveries from settled Europeans in Australia were underplayed or ignored.

For example, significant members of the Australian community had already reported that platypuses lay eggs decades before Caldwell arrived. In 1825, New South Wales Supreme Court Judge Barron Field made the following throwaway comment about the apparent oddness of Australia: 'But this is New Holland . . . where the swans are black and the eagles white . . . where the mole . . . lays eggs, and has a duck's bill.'[23] Likewise, Sir John Jamison – former naval surgeon under Lord Nelson at Trafalgar – had written to the Linnean Society in 1816 with the unevidenced news that 'the female is oviparous, and lives in burrows in the ground'.[24] While several nineteenth-century writers mention this report in passing, few of them take him at his word.

It seems that Europe would only believe this apparently unlikely zoological fact if they heard it from one of their own.

*

Caldwell and Balfour were Cambridge men. So, soon after I joined the University Museum of Zoology, I asked our collections manager, Mathew Lowe, whether we still had Caldwell's Australian specimens. I had a list of what he had shown the Royal Society of New South Wales and wondered if any of those made it home with him – and indeed whether it included any of the groundbreaking specimens from August 1884. With over 1,300 echidnas killed for him in 1885, I imagined there would be an entire aisle in a storeroom to house the 'Caldwell Collection'.

But Matt had never heard of him, which immediately

suggested Caldwell was not a major collector for the museum. Upon checking, there were no specimens attributed to him on our collections database: 'Drew an utter blank from all angles on Caldwell, I'm afraid.'

That didn't discourage me. There isn't a natural history museum on Earth that actually knows what's in its own collections. With hundreds of collectors adding two million specimens over more than two centuries, it is effectively impossible to imagine a complete list of holdings for a museum as large as ours in Cambridge. In university collections the situation is typically even harder, as what belongs to the museum and what belongs to individual researchers working near the museum is often a grey area, and the result of a person's entire career might have been suddenly landed on the museum after their retirement or death, with no real capacity to document it, sometimes until decades later.

'Please keep an eye out, will you?' I asked Matt. Every museum worker has a list of things they are 'keeping an eye out for', gathered over the course of their entire careers from countless colleagues and visiting researchers, each with their own pet project. It rarely works out, but when it does, this is a key route by which museums get to learn about the value of specific items in their collections – when other people tell them about the importance of what they've got (there's no way we could discover every specimen's history on our own).

While despite all that I optimistically pictured Matt coming into my office ten years hence, holding a ceramic jar filled with echidnas, I didn't think it would actually happen.*

It took him less than three months. It wasn't a ceramic jar, however, and it's not exactly an aisle full of specimens. Among a small tub of jars that were found at the back of one of the teaching labs in the 1990s, and another that was transferred

---

* And of course, proper museum practice would have involved Matt taking me to the jar, not bringing the jar to me.

from the anatomy department around a similar time, were twelve miniature vials, less than 10 centimetres (4 inches) tall. Four of them are labelled as belonging to Caldwell (the others were all infant marsupials and monotremes, so possibly Caldwell's too, but we'll probably never know).

We already know that Caldwell was not sentimental about the biology he was doing and appeared to have little interest in the animals beyond their embryology. Nonetheless, I had been clinging to the possibility that Caldwell had at least kept some complete animals. If he did, they have not survived. Perhaps it's not surprising that these four remaining links to his monumental discoveries are not beautifully prepared, whole adult specimens.

Naturally, embryological specimens are not much to look at. To start with, they are small. Much of Caldwell's original material will have been prepared as microscope slides (possibly with the Caldwell Automatic Microtome). But what we uncovered in our stores were two echidna-pouch young (one headless – presumably having been sliced for slides), one platypus uterus sharing its vial with a baby possum and four unidentified infant marsupials. Although I found a couple more babies the following year, I live in hope that more of the many platypuses or 1,300–1,400 echidnas the team of Indigenous collectors amassed will pop up in the collection over time, but I feel privileged to work alongside these few remnants of their efforts.

Prior to joining Cambridge, I led the Grant Museum of Zoology at University College London. Its progressive founder Robert Edmond Grant prepared specimens of platypus reproductive systems that were some of my favourite objects there because of their importance to platypus lore. Early in the debates, Grant used them as evidence to support the theory of egg-laying in platypuses. He had a career-long feud with Richard Owen that covered many issues. In the pages of journals, and the official and unofficial chambers of learned societies, they fought battles big and small. Topics included the identification of individual fossils, whether animals evolve and whether monotremes lay eggs. Eventually Owen – despite being on the losing side of several of these debates – played a significant role in the destruction of Robert Grant's career. Owen is the reason that you've probably never heard of Robert Grant.

He really ought to be more famous than he is. Who knows what biology would look like without Grant's influence? Starting in the 1820s, he was most probably the first person to teach evolution at an English university; two of his students would become directors of the Natural History Museum – or British Museum (Natural History), as it was known then; he collaborated closely with some of Europe's leading anatomists (particularly the liberally minded ones); and as his mentor at the University of Edinburgh, Grant taught a young Charles Darwin about evolution, which of course changed the world.

When it comes to the platypus, Grant thought he'd nailed it – fifty-five years before Caldwell. Arguably Grant's most influential friend was Étienne Geoffroy Saint-Hilaire – who had come up with the name *monotremes* for the group – another progressive and one of France's most eminent scientists, based at the Museum National d'Histoire Naturelle in Paris. Geoffroy was one of those leading the debate on the side of egg-laying in monotremes in the early decades of the nineteenth century. He thought that they represented an intermediate form between reptiles and mammals, and therefore bolstered the evidence for evolution.[25] If Owen was the poster boy for internal hatching, Geoffroy was the champion for egg-laying. If Grant could help prove that monotremes laid eggs, he would be able to get one up on Owen.

Grant wrote to Geoffroy in September 1829 with big news. He had just examined some platypus eggs brought to London by a natural history collector who had found them in a hollow on a sandbank, from where he had watched a platypus enter the water.

You already know that this does not have a happy ending, otherwise I wouldn't have spent all that time talking about

Caldwell. Although Grant could not convince the collector to sell him the eggs (a source of great regret for me when I was at the Grant Museum), he described the size and structure of the shell and provided a detailed illustration. Without explanation, Owen didn't accept Grant's findings,[26] but Geoffroy took Grant at his word – although the large size that Grant described (3.5 centimetres/1½ inches long and 1.9 centimetres/¾ inch wide) caused Geoffroy some trouble to explain how that would pass through a platypus. Based on more recent analysis, we now know that what Grant had seen were the eggs of a freshwater turtle.

In the same way that the socially conservative Owen clung on to his view that platypuses produced milk but could not lay eggs, his liberal rival Geoffroy staunchly defended his opinion that they laid eggs but could not produce milk. On hearing the report of Lauderdale Maule's still-warm carcass oozing milk he purportedly said, 'If these glands produce milk, let us see the butter'.[27] Both men were steadfast in their stubbornness, with a keen eye on the political interpretations of their positions on the platypus.

It is through specimens like Caldwell's embryology material and Grant's reproductive system that we see how museum objects can be imbued with such historical significance. They demonstrate how the controversy over platypus egg-laying was a key battleground for pre-Darwinian evolutionists like Grant and conservatives like Owen. Not only that, but preserved specimens like these – along with colonists' descriptions and illustrations – were the key ways by which people in Europe came to know Australia's animals.

6

Terrible with Names

From their first arrival, Europeans began describing Australia's fauna – and describing it badly. Taxonomy – the scientific framework on which animal names are built – made matters even worse. The names ascribed to species can give us clues about how they were considered by the people who named them, as can the scientific (or otherwise) accounts that describe the newly known animals. Often, they point to hierarchical language, with some kinds of animals being given grander titles than others, based more on societal preconceptions than any objective considerations. And very often, the names are just plain wrong.

Naturally, the animal that the British called 'platypus' already had plenty of Indigenous names before Europeans arrived with their hierarchical naming systems. Ones that are widely quoted in English-language publications today are *mallangong* and *tambreet*, which were first reported by George Bennett in 1835 as being used in regions of what is now southern and central New South Wales.[1] When I put a call out, the people of Twitter provided these Indigenous names for the animal:*

---

* With great thanks to all those who contributed.

Group, language or language group	Word for the animal
Bundjalung	ḏanbaŋ
Djab wurrung	Mirwil
Djadjala	Mabial
Dyirbal	gugula
Gamilaraay	buubumurr
Kuurn Kopan Noot	allertil
Ngarigo	djamalan
Ngunnawal	malungan
Peek Whurrong	torron'gil
Wemba-Wemba	mapiyal
Wiradjuri	Biladhurang, wamul, djiimalung, dungindany
Wurundjeri Woiwurrung	dalai-wurrung (which I'm told means 'fat-lips')
Yorta Yorta	wannagapippua

When the British arrived, they first considered platypuses to be a species of mole and called them duck-moles. I rather like that. Platypus, by contrast, simply means 'flat foot' in Greek, which is obviously not the most noticeable thing about them.

Platypus as a common name is a hangover from their first scientific name, given by George Shaw in 1799, which was *Platypus anatinus* (literally 'flat-footed duck-like', even though ducks' feet are significantly flatter than platypuses'). However, it turned out that the *Platypus* designation already belonged to a wood-boring beetle, and because no two species can have the same scientific name, a new one was required. As it happened, in his accounts of the dried specimens Joseph Banks had sent him, the German anatomist Johann Blumenbach had proposed the name *Ornithorhynchus paradoxus* ('bird-nosed paradox' or 'bird-nosed strange thing') a year later, unaware that Shaw had already named them *Platypus anatinus*.

Normally, when one animal has been given two different names, the strict rules of taxonomy require us to only use the one that was published first and ignore all others (even if we prefer the second one). However, in this case the first part of the name used earlier by Shaw, the genus name – *Platypus* – was invalid as it was already in use by the beetle. So, Blumenbach's genus name – *Ornithorhynchus* – was combined with the second part of Shaw's original name, *anatinus*. As such, the platypus officially became *Ornithorhynchus anatinus*, meaning 'bird-nosed duck-like'.

They have since been described using better names, but the original 'platypus' has stuck in everyday English as a common name. And as is often the case when two names appear in the literature close together in time, many people continued to use Blumenbach's *Ornithorhynchus paradoxus* for many decades afterwards, even though it was invalid. Presumably it became popular as it perfectly summed up the perplexing nature of the platypus.

Other non-Australian languages have different words for the platypus. Many are thoroughly non-specific names that could apply to all birds, squids, octopuses, turtles and many dinosaurs, with names that mean 'bird beak', 'beak-animal' or 'duck-nose'. In Iceland they are called 'wide-nose'. The Croatians introduce a value judgment by calling them 'strange beak'. And some names are improvements on the English name: the Vietnamese word translates to 'interesting bill of duck' and in Kiswahili it's 'bird-beast'. Hungarian is surely the most accurate: 'mammal with the beak of a duck'.

Again, I asked people on Twitter to share their languages' names for what in English is known as the platypus, along with the direct English translation of the words' constituent parts. I've trusted what I was told here, because verifying the etymology of these isn't always easy:

Language	Word for the animal	Direct English translation
Afrikaans	Eendbekdier	duck-mouth-animal
Armenian	բադակտուց (badaktuts)	duckbill
Bahasa Indonesia	Platipus	platypus (transliteration)
Bulgarian	Птицечовка (pticechovka)	bird-beak
Chinese	鴨嘴獸 (yazuishou)	duck mouth beast
Croatian	čudnovati kljunaš	strange beak
Czech	Ptakopysk	bird-lip
Danish	Næbdyr	beak-animal
Dutch	Vogelbekdier	bird-beak-animal
Finnish	Vesinokkaeläin	water bill-animal
French	Ornithorynque	bird-beak
German	Schnabeltier	beak-animal
Greek	Ορνιθόρυγχος (ornithorynchos)	bird-beak
Hebrew	זוורב (barvezan)	duck-like
Hungarian	kacsacsőrű emlős	mammal with the beak of a duck
Icelandic	breiðnefja	wide nose
Irish	platapas	platypus (transliteration)
Italian	ornitorinco	bird-beak
Japanese	鴨嘴 (kamo-no-hashi)	duck's bill
Lithuanian	ančiasnapis	duck beaked one
Norwegian	nebbdyr	beak animal

Language	Word for the animal	Direct English translation
Persian	یکدرا کون (nokordaki)	duck beak
Polish	dziobak australijski	Australian beakish
Romanian	Ornitorinc	bird-beak
Russian	утконос (utkonos)	ducknose
Scots	deuk-nebbit platypus or deukmowdie	duck-beaked platypus or duck-mole
Scottish Gaelic	platapas gob-tunnaige	duck-beaked platypus
Slovenian	kljunaš	beak-er
Spanish	ornitorrinco	birdbeak
Swahili	kinyamadege or domobata	beast-bird or beak-duck
Swedish	näbbdjur	bill-animal
Thai	ตุ่นปากเป็ด	mole-mouth-duck
Ukrainian	kachkodz'ob	duckbill
Vietnamese	Thú mỏ vịt	interesting bill of duck
Welsh	hwyatbig / platypws	duck-beak / platypus (transliteration)

On the topic of language, it's also worth mentioning that the plural of platypus isn't platypi (the same is true for octopus). This is because '-pus' is Greek in origin, and therefore doesn't pluralise to '-pi' as it would if it were Latin. If you wanted to be extremely pretentious and stick with Greek grammar, it would become platypodes. But let's just stick with English and use platypuses.

*

Looking at how the language of taxonomy is applied to other Australian mammals, one also notices that marsupials are frequently given the short straw by being denied an identity of their own. For example, the thylacine is commonly called the 'Tasmanian tiger' due to its stripes, but that demotes it by suggesting it is merely a sub-set of tigers. Like all marsupials, thylacines are as closely related to tigers as they are to elephants or humans. I would prefer we avoided that name and stick to the one they can call their own.

I will introduce the different kinds of marsupials properly in the next chapter, but when it comes to their scientific titles, some do have names that set them apart for their uniqueness: *Macropus*, the genus containing kangaroos and most wallabies, means 'big foot'; *Petaurus*, the name for some of the gliders, and *Acrobates*, the feather-tailed gliders, have names that reference their agility, meaning 'rope dancer' and 'acrobat' respectively; the quolls are called *Dasyurus* for their 'hairy tails', and Tasmanian devils are called *Sarcophilus*: 'flesh-lover'. Other scientific names are derived from Indigenous names, such as *Bettongia* for bettongs, *Potorous* for potoroos (both petite, hopping relatives of kangaroos), and *Wyulda*, which is the scaly-tailed possum's genus but is actually an Indigenous word for brush-tailed possums that has been incorrectly applied. However, very many are simply named after animals that Europeans are familiar with. For example, *Dendrolagus* – the tree-kangaroos – means 'tree-hare'; among the small carnivores, *Antechinus* means 'hedgehog equivalent' and dunnarts are called *Sminthopsis,* which means 'appearance of a mouse'.

Often, marsupial common and scientific names are just derivatives of placental mammal names formed by adding versions of a word that means 'pouched' (*phasco, thyla* and *pera*):

The koala's scientific name is *Phascolarctos*: 'pouched bear'.
Some bandicoots have the scientific name *Perameles*:
    'pouched badger'.

Thylacines are named *Thylacinus cynocephalus*: 'pouch-like dog-head'.*

Phascogale – the squirrel-sized predatory marsupials – means 'pouched weasel'.

The name echymipera (also called spiny bandicoots) means 'pouched hedgehog'.

Dasyures are New Guinean carnivores with the scientific name *Phascolosorex*: 'pouched shrew'.

It's as if these marsupials are only second-rate cover versions. Worse, the names are often wholly inappropriate: they do not even fit the descriptions. For example, the scientific name for pademelons – cat-sized relatives of wallabies – is *Thylogale*, which means 'pouched weasel'. How, you might ask, does a 5-kilo (11-pound) marsupial that hops on its hindlegs resemble a 100-gram (3½-ounce) tube-shaped weasel? In fact, many marsupials are bizarrely named after weasels, when only the phascogales have a shape that could be argued to be weasel-esque (and even then, phascogales' bottle-brush tails are a significant point of difference):

Rock-wallabies are *Petrogale*: 'rock-weasel'.

Planigale (the smallest marsupial carnivores) means 'flat weasel'.

The nail-tailed wallabies are *Onychogalea*: 'claw-weasel'.

---

* *Thylacinus* is almost universally said to mean 'pouched dog', from the Greek *thylakos* (pouch) and *kyon* (dog); and so *Thylacinus cynocephalus* would mean 'pouched-dog dog-head' (Strahan & Conder, 2007). However, thylacine-researcher Douglass Rovinsky pointed out to me that in other names, '-inus' means 'like' – for example, *anatinus* means 'duck-like' in the platypus' name. Additionally, we could think of no other times when a reference to dogs was spelt with an I, as in –cinus, rather than a Y, as in *cynocephalus*. I have checked the original French publication that names the animal *Thylacinus* (Temminck, 1824) and there is no etymology given for the name. I now believe that *Thylacinus cynocephalus* directly translates to 'pouch-like dog-head'. Temminck had seen no female specimens, and mentions the 'sac' into which the male's scrotum fits, so perhaps that is the pouch the name refers to, but this is just my conjecture.

Marsupial common names can be thoroughly nonsensical. Spotted-tailed quolls were called 'tiger cats' by colonists – a name that has stuck – even though they are half the size of a cat, are spotty not stripy and are marsupials. I think quolls are far more like spotted mongooses or members of the polecat and marten family than cats (and indeed the scientific name for eastern quolls is *Dasyurus viverrinus*, which means 'ferret-like hairy-tail'), yet they are collectively known as 'native cats'.

Even the group names are judgmental. The scientific name for the placentals is Eutheria, or 'true beasts', whereas marsupials are Metatheria, or 'altered beasts'. The literature is littered with phrases that imply Australian mammals are inferior in comparison to placentals. Celebrated mammalogist Richard Lydekker, who published extensively on the Natural History Museum's collections, regularly drops in terms like 'True Mammals' and 'ordinary mammals' in contrast to the monotremes.[2] Richard Owen says that marsupials and monotremes 'may be regarded as an aberrant group of Mammalia'.[3]

Names matter in other ways, too. Today, some animals (in Australia and elsewhere) have been rebranded to make the public care more about them. Marketing strategies are essentially being applied in conservation by changing names that don't rate well among the public or have negative connotations. For Australian mammals, this often involves removing the word 'rat' or 'common' from their names (some animals called 'common' aren't actually that common, and it gives the impression that they are not priorities for conservation). The prehensile-tailed rat has become the 'tree mouse'; the water rat has become 'rakali' (one of its Indigenous names); common wombats have become 'bare-nosed wombats'.[4] The list goes on.

As an aside, I've noticed that when Australian mammal species are named after a feature that only one sex possesses, they are almost universally named after an aspect of female anatomy. For one thing, the entire marsupial group is named

This 1864 print from a French magazine depicts a group of hairy-nosed wombats, with a focus on their powerful backsides. It was drawn by the artist Charles Jacque from live wombats in the Jardin d'Acclimatation park in the Bois de Boulogne in Paris.[5]

for the female's pouch: *marsupium* is Latin for 'purse'. And across the names of different marsupials we see references to the pouch come up time and again. As such, at least half of all the members of each species (the males) do not have the features by which their species is described.[6]

*

Perhaps it's time to properly introduce my second-fondest zoological love – wombats. They, too, have been subject to unfortunate misnaming. One of their outdated names was *Phascolomys*, which means 'pouched mouse'. Wombats can weigh over 30 kilograms (66 pounds), so such a comparison is absurd. Fortunately, the valid scientific name of a common wombat is *Vombatus*, which is a Latinised version of a Darug name for the species: wombat or wombach (there is no W in Latin).

I could recommend these cuboidal herbivorous marsupials simply on the basis of some of their impressive adaptations.

For example, wombats are able to deal with the tough plants that form their diet because of their ever-growing open-rooted teeth. Our teeth have closed roots, so once we've worn them down or damaged them, they don't grow back. The surfaces of wombats' teeth are constantly being ground away by the plants they eat, and also because they sometimes gnaw through harder areas of soil when digging (or to gain extra minerals). No matter, though, as their teeth regrow from the gum end at the same rate.

Another wonderful wombat ability is that if a predator chases them into one of their extensive burrow systems, the solid, reinforced shape of their backsides – sloping to form a noticeable overhang – enables them to crush the heads of their would-be attackers against the roof of the tunnel with their bums. Compressed fox and dog skulls have been found in wombat warrens.[7]

Perhaps the most famous fact about wombats, however, is that they poo cubes. Each night, one common wombat can produce as many as 100 almost perfect 2-centimetre (1-inch) cubic scats. It was only recently discovered how this is biologically possible – in 2018, Patricia Yang announced that the last section of a wombat's intestines does not stretch uniformly. There are alternating bands of stiff and more flexible tissue around their circumference. As the intestine fills and expands, the poos are shaped into cubes. This is interesting not only for its own sake, but because it represents a new way for engineers to create a cube, potentially having an implication for manufacturing.[8] And then, in 2023, researchers discovered how wombat poo breaks into regular cubes. This work was inspired by the way volcanic lava can cool to form perfectly hexagonal columns, like in the Giant's Causeway in Ireland – it's all about how the intestines extract water from the stool. While the different muscle bands are forming the edges, wombat guts dry their poo at such a rate that it cracks into blocks as tall as they are wide.[9]

That explains how they do it, but to answer *why* we have to look at *where*: wombats do their pooing on top of prominent objects in their habitats such as rocks, logs, tussocks or any noticeable feature (wombats often get blamed for lots of destructive digging because people find their poos on top of the freshly dug soil, but often this is simply a consequence of their habit for pooing on novel noticeable landmarks – feral hogs are usually the true culprits). Defecating on objects raises the poo up off the ground, which increases the spreading of its scent, and also moves it nearer nose height. In her wonderful book *The Wombat*, Barbara Triggs suggests that this may help wombats create a mental map of their home range, and act as signals to other local wombats. Until the age of about two years, wombatlets that have left the pouch hide in bushes while pooing, and only switch to this extremely visible behaviour as they are establishing their home range and approaching sexual maturity.[10] The importance of the poos being cubic is that they are less likely to roll off these prominent perches.

Because wombat pouches face backwards (which is beneficial as it means they don't fill with soil while the mothers are digging, and is also presumably more convenient when a large wombatlet is climbing in and out of the pouch), it does mean that when wombatlets reach the point that they stick their head out, they can find themselves face down in their mum's poo, or indeed any other poo that she walks past (because wombats make so much poo, and particularly as in drier and colder places it can break down very slowly, wombat habitats are often heavily strewn with these cubes). This may be a good thing, however, as it is believed that wombatlets need to eat poo in order to gain the gut bacteria required to break down their plant diet.

When it comes to the young wombatlets themselves pooing, the situation is a little less clear. In other marsupials, the mother reaches her head into the pouch and licks away any faeces and urine from the baby's cloaca. However, it would be very difficult for a female wombat to reach her pouch with her mouth because of its orientation – particularly if the wombatlet were large (they can approach a hefty 4 kilograms/9 pounds and still be carried in the pouch at 8–10 months old). Just as baby wombats stick their heads out and feed on grass with the rest of their body held safely in the pouch, there have been observations of them sticking their tails out and doing the opposite.[11]

I could continue this list of excellent wombat skills and provide some scientific reasoning for why I think they are top;* however, I have to admit that my attraction to them is not very scientific. The truth is, I am most drawn to animals that waddle – animals for which one could imagine an internal monologue or theme tune of slow, bumbling, rhythmic muttering as they move. If such an animal had a musical theme in Prokofiev's *Peter and the Wolf*, for example, it would go 'bomp-te-bomb, bombety-bomp-te-bomb' over and over as it walked. That is what I imagine wombats humming to themselves as they walk, and that's why I like them.

As Ivan Smith puts it in his heart-wrenching short story of the impact of bushfires on wildlife, *The Death of a Wombat*:

> The wombat comes from a pleasant family, fussy and gentle, slowminded, and polite. He is a close cousin of the koala bear, who took to the trees a long time back to get away from it all. The wombat has a short snub-nose and short, stubbed legs and a short-range mental life. He is a ponderous plump of meat with a lurching walk.
>
> Everything likes a waddler. The wombat lurches on, slowly minding his own business. There are bits of bark to find,

---

* 'Top' was actually the name of the pet wombat that the Pre-Raphaelite painter Dante Gabriel Rossetti kept in his London home in 1869. It was mockingly named after his artist friend William Morris, whose nickname was 'Topsy'. This was clearly intended as a slight against Morris, as Rossetti was utterly obsessed with Morris's wife, Jane (one of his drawings of Jane has Top at her feet, both with angelic halos over their heads) (Simons, 2008).

and things to visit. And everything likes a waddle and crump, and slowly home to dinner on time, and gentle doze in a well-made hole, and early thoughtless yawns, and waddle and crump again.[12]

Echidnas do the same thing (in my mind, at least), and so do bears. The comparison to bears is a rare occasion where an Australian mammal's scientific name actually does make some sense. The common wombat's full name is *Vombatus ursinus*, meaning 'bear-like wombat'. There is no denying that wombats bear a striking resemblance to bears, and not just in the way they waddle.

I should also point out that despite the impression of being 'slowminded', as Smith put it, there is no reason to think that wombats are unintelligent. On the contrary, they have the largest brains of any marsupial, and wombatlets engage in complex play behaviour. I like to think that the apparently ponderous pace with which they approach life is simply a manifestation of the fact that they know that they can get themselves out of most situations. It's not dim-wittedness, it's the nonchalant swagger of self-confidence.

Although much of wombat life is carried out at a relaxed waddling pace, when they want to, they can sprint at incredible speeds. I've been in a vehicle and observed wombats running through the bush nearby and maintaining pace with the car. Using back-of-the-envelope calculations, Tasmanian devil biologist Rodrigo Hamede Ross and I have calculated that wombats could outpace Usain Bolt over short distances. In one of those elaborate schemes that one tends to develop on field trips when isolated with a very small group of people in the wilderness for an extended period, we've discussed at length the idea of staging a race between a wombat and Bolt. Of course, we know that whatever the outcome, this is a race that Bolt could only lose (if he outpaced a wombat, no one would think it were a big deal, but if the wombat crossed the line first, well, the world's fastest man lost his crown to a dumpy, lumbering medium-sized mammal).

Nonetheless, Usain, if you're reading this, think of what such a race would do for the profile of this animal.

It's not just their speed or their ability to crush a predator's skull with their backsides that makes wombats appear untouchable – their teeth can also be vicious weapons. They are extremely sharp, and the jaws incredibly strong, which means that wombat bites can cause serious injury (something that they can do to each other when wombat battles erupt, presumably over the ownership of home ranges). What's more, their skulls are solidly fused, and with their muscular strength they can use their heads as battering rams. This is typically employed to move large and heavy objects out of their paths, but it is an equally useful skill when tackling a would-be predator. Don't be fooled by a wombat's gentle appearance.

*

Europe's relationship with marsupials did not begin in Australia. They had been known in Europe since Vicente Yáñez Pinzón sailed with Christopher Columbus as captain of the *Niña* before recrossing the Atlantic and finding what we now call opossums in Brazil in 1499. According to a 1671 translation of his description of the animal, Pinzón was not impressed. Nevertheless, in this, the first known European encounter with a marsupial, he managed to report the key feature for which the group is famous. He saw:

> as strange a Monster, the foremost part resembling a Fox, the hinder a Monkey, the feet were like a Mans, with Ears like an Owl; under whose Belly hung a great Bag, in which it carry'd the Young, which they drop not, nor forsake till they can feed themselves.[13]

Despite his apparent revulsion, in 1500 Pinzón brought the animal back alive to Spain and presented her to King Ferdinand

and Queen Isabella, who stuck her hand in the pouch to find the shrivelled babies that had died during the journey. In 1563, Conrad Gessner illustrated an opossum, making the animals widely accessible.[14]

This means that we are now into our sixth century of knowing marsupials in Europe. But looking back at the accounts of some of the earliest meetings between Europeans and Australian marsupials, we get a sense either that the writers' descriptive abilities were so poor that no one who followed in their footsteps would be able to recognise the animals they wrote about, or that the appalling value judgments they introduced would cloud the way the rest of the world thought about the animals.

Just in case it needs saying, Captain Cook did not discover Australia – this is a widespread but utterly false British colonial narrative. The clearest counter to that suggestion is, of course, that Indigenous Australians had been there for at least 60,000 years by the time James Cook and Joseph Banks arrived on the east coast in 1770.

And other Europeans had arrived hundreds of years before them. There are sixteenth-century maps (including the Dieppe maps) that appear to include Australia's northern shore in rough position relative to New Guinea, including one from 1566 (by Frenchman Nicolas Desliens) and one from 1578 (by Flemish explorers Gerard and Cornelis de Jode), although these are contested, and there is no evidence that any of their ships landed. One theory for why any such encounters would have been kept secret is that they were made by Portuguese sailors who were not supposed to be in that part of the world, as any discoveries there technically belonged to Spain, under a treaty drawn up by the pope.[15]

Some believe that sailors from the Arabian Peninsula, China and Malaysia were exploring the area long before that. In any case, the first confirmed European contact with Australia was in 1606, when Willem Janszoon, sailing aboard the *Duyfken*

('Little Dove'), charted part of the western coast of Cape York (the far northeast of the country). He believed it to be an extension of New Guinea, having missed the existence of the Torres Strait. This treacherous body of water was navigated that very same year, by Spaniard Luis Vaez de Torres – a landmark sailing that would be kept secret for over 150 years (nations often guarded their navigational discoveries closely, to maintain a competitive advantage over their rivals). Indeed, no European would cross the strait again until Cook did so in 1770, ending speculation that New Guinea and what we call Australia were one landmass.

The country's original European name was 'New Holland', and in the 1600s Dutch mariners – as well as occasional visits from Spanish vessels – routinely travelled the western shores on trade routes between South Africa's Cape of Good Hope and the Dutch colonies in Java. Many of these sailors' encounters with the country were accidental – the Brouwer Route, for example, was a newly discovered means of reaching Java far quicker than skirting the East African coast before crossing the Indian Ocean. It involved sailing south from the cape until the 'Roaring Forties' (powerful westerly winds) took hold and pushed the vessel eastwards, and then turning north after 6,000 kilometres (3,700 miles). Naturally, at these distances it's easy to overshoot. Occasionally these errors led to new opportunities to chart unknown lands (one voyage in 1626/7 was pushed 1,500 kilometres (930 miles) further along Australia's south coast), occasionally they led to shipwreck, and occasionally both.

As knowledge of the size of the continent grew, the imperial establishment began to speculate about what could be exploited there for gain. The instructions given to one voyage, by the ruthless governor-general of the Dutch East Indies, Jan Pieterszoon Coen, were to survey New Holland's west coast for valuable resources and trade opportunities, to take possession for the Dutch Empire and:

In places where you meet with natives, you will either by adroit management or by other means endeavour to get hold of a number of full-grown persons, or better still, of boys and girls, to the end that the latter may be brought up here to be turned to useful purpose.[16]

Fortunately, those ships never reached New Holland in order to enslave Aboriginal people. And subsequent voyages didn't find anything the Dutch considered valuable enough to justify the hardships of visiting the parts of Australia they had seen. The sailing was extremely dangerous, fresh water was thoroughly hard to come by and they failed to find evidence of spices, gold, fruits or minerals to trade. Eventually the shareholders of the Dutch East India Company would give instructions to effectively stop exploring.

Nonetheless, by as early as the 1640s, much of Australia's outline had been joined up on European maps by the Dutch, including the Gulf of Carpentaria and Arnhem Land in the north, all of Western Australia's coast, and as far east as the Great Australian Bight on the south coast. Tasmania was added to that list by Abel Tasman in 1642.[17]

On each of their voyages, early European explorers consistently failed to find suitable words to describe the animals they encountered. Just as Pinzón had done, following that first European meeting with marsupials, they made inappropriate and inaccurate comparisons to well-known animals, rather than giving an account of the animals' individuality.

Cook and Banks found eastern grey kangaroos on their visit in 1770. Other species of hopping marsupials had already been described on earlier European voyages in and around Australia, but unfortunately, the descriptions of these are largely so invalid that each subsequent explorer – despite having read the preceding voyages' published accounts – failed to recognise they were seeing similar animals to those that earlier expeditions had found.

The first European representation of an Australasian animal may be an illustration on the title page of Cornelis de Jode's 1593 *Speculum orbis terrae* (adding weight to the possibility of pre-1606 European encounters). Although the drawings are not explained, the page has animals in each of its four corners, and it could be interpreted as each of those animals representing a quadrant of the world map: top left is a horse, for Europe; top right is a camel, for Asia; bottom left is a lion, for Africa. The animal at bottom right, in the relative position of Australia, is not easily recognisable. It is half lying on its belly and half sitting on its legs, which are muscular, roughly of equal length and with forepaws resembling hands. It has a long neck and a camel-like head. In those respects (except the hands) it resembles any large, generic four-legged animal. However, the remarkable thing about it is that at the base of its neck, between its front legs, is a large sack, which holds two equally sized babies.[18] Could this be the first Western depiction of an Australian animal – and indeed a kangaroo – potentially based on sailors' descriptions after visiting the north coast?

As I say, no words accompany that illustration, whatever it depicts. The earliest known written account of an Australasian (but not Australian) mammal is from 1606, when Don Diego de Prado y Tovar, sailing with Torres, described what was probably a dusky pademelon (a wallaby relative), which he encountered on the southeast coast of New Guinea. He wrote that it was:

> In the shape of a dog smaller than a greyhound, with a bare and scaly tail like that of the snake, and his testicles hang from a nerve like a thin cord; they say that it was a castor [referring to a beaver, perhaps because of its teeth, scaly tail or odour], we ate it and it was like venison.[19]

And there begins the European habit of describing Australasian mammals by using inappropriate comparisons to familiar animals from other parts of the world. In this short note, the small hopping marsupial is compared to a dog (twice), a snake, a beaver and a deer. There is no mention of any of its most notable features (aside from the novelty of the shape of the marsupial scrotum, which typically resemble garlic bulbs or squat figs), such as its massive powerful hindlegs, extraordinarily long feet, small front legs, back-heavy proportioning, prominent ears or long face. I appreciate that it's only natural to describe new species by comparing them to familiar ones. The trouble is, however, that it's only useful if it's done in a way that makes clear the points of difference – what are the animals' defining features?

The next written account to appear, describing an animal on Australian soil, is a lot better in that it is possible to hold an image of the animal it relates to – a tammar wallaby – in your head without it disintegrating as you read. It was written in astonishing and horrendous circumstances in 1629, by Dutch merchant Francisco Pelsaert, after meeting the wallabies in the Houtman Abrolhos, a chain of islands 80 kilometres (50 miles) west of modern-day Geraldton, Western Australia. But quite a lot happened before that.

His ship, the *Batavia*, was carrying the equivalent of nearly $100 million in cash – intended to finance Jan Pieterszoon Coen's expansive operations in the Dutch colony of Batavia on Java – when it wrecked on a shoal. The ship's longboat and yawl managed to ferry 240 passengers to a nearby island, while the crew worked on the wreck to salvage the precious cargo. Pelsaert – along with forty-eight of the passengers – then spent ten days searching nearby land with the longboat, failing to find water for the stranded survivors. After picking up a measly 80 litres (21 gallons), they were pushed back out to sea by a storm (which, as it happens, delivered enough rain to keep those left on the island alive), and decided to make a run for Batavia in their tiny open vessel. They reached the Javan coast ten days later and eventually arrived in the port a month after the *Batavia* wrecked.

Naturally, Coen was eager to salvage his treasury (and the missing people), and so sent Pelsaert back in a yacht. Things had not gone well since he left. Upon Pelsaert's return to the island, a pack of 'scoundrels' from among the survivors attempted to take the rescue yacht and escape. Pelsaert fought them back and managed to take them prisoner. He learned that the castaway camp had descended into a bloodbath, in which 125 men, women and children had been killed by these mutinous members of the crew.[20]

In the two months it took him to salvage the wreck, Pelsaert also tried and hanged the perpetrators (the first European-style trial on Australian soil). And after all that, he still had the presence of mind to describe the wildlife, in what would be the first European account of an Australian mammal:

> on these islands there are large numbers of cats, which are creatures of miraculous form, as big as a hare, the head is similar to that of a civet cat, the fore-paws are very short, about a finger long. Whereon it has five small nails or small fingers, as an ape's fore-paw, and the 2 hind legs are at least half an ell long [roughly 30 centimetres or 1 foot], they run on the flat of the joint of the leg, so that they are not quick in running. The tail is very long, the same as a meerkat, if they are going to eat they sit on their hind legs and take the food with the fore-paws and eat exactly the same as the squirrels or apes do. Their generation or procreation is very miraculous, yea, worthy to note, under the belly the females have a pouch into which one can put a hand, and in that she has her nipples, we have discovered that in there their young grows with the nipple in its mouth, and have found lying in it [the pouch], some which were only as large as a bean, but found the limbs of the small beast to be entirely in proportion, so that it is certain that they grow there at the nipple of the [mothers] and draw the food out of it until

they are big and are able to run. Even though when they are very big they still creep into the pouch when chased and the mother runs off with them.[21]

You'll notice that, far from Pinzón's 'Monster', Pelsaert used the word miraculous twice to describe what he saw. It's particularly remarkable, as among all the bloodshed, the survivors had accurately observed details of their reproduction and maternal care, which are indeed 'very miraculous'. Some translations of this journal entry have it that the babies 'grow out of the nipple', rather than 'grow there at the nipple', which started a long-held belief that this is how kangaroos reproduce.

After Pelsaert met the tammar wallabies, two Dutch explorers encountered the now-famous quokkas on Rottnest Island (off modern-day Perth). First, in 1658, Samuel Volkersen described 'a wild cat, resembling a civet-cat, but with browner hair'.[22] Then, in 1696, Willem de Vlamingh (on a rescue mission to look for a Dutch East India Company vessel that had gone missing between the Cape of Good Hope and Java) gave Rottnest its name (meaning 'rat's nest'), saying: 'of animals there is nothing but bush rats'.* I find a couple of things about these reports interesting. The first is that both Pelsaert and Volkersen landed on civets as their placental comparator for two quite different hopping marsupials (tammar wallabies and quokkas). Civets are often called civet cats but are not cats – they are fairly elongated carnivorans with mid-length legs and often blotched or spotted fur – and don't resemble either of those marsupials in any obvious way. This could suggest that the second report was influenced by the first, but if so, why not say as much? The second thing is that a quokka is

---

* Vlamingh's crew also have the distinction of being the first Europeans to set eyes on a dingo, when they landed on the Australian mainland in 1697 near what is now Perth. They are also believed to be the first Europeans to encounter Australia's black swans, and gave the Swan River its name in their honour. (Whitley, 1970)

well over ten times bigger than a rat, and is in no way rat-like. Or cat-like, for that matter. Nonetheless, the Dutch politician Nicolaes Witsen, who sponsored that 1696 voyage, reported:

> [Vlamingh] found no people but a large number of rats, nearly as big as cats, which had a pouch below their throat into which one could put one's hand, without being able to understand to what end nature had created an animal like this: as soon as it had been shot dead, this animal smelled terribly, so that the skins were not taken along.[23]

Reading this now, it's hard to predict how much the understanding of zoology would have been brought forward if the sailors had had less sensitive noses and had returned to Europe with specimens. For what it's worth, my experience is that dead quokkas smell no more or less than any of their marsupial relatives, but I can think of plenty of mammals which smell infinitely worse dead than a quokka. Rottnest is now one of Australia's tourism hotspots, mostly because of the accessibility of the quokkas. It is genuinely impossible to go there and fail to see one. I hesitate to use the word 'cute', but these miniature kangaroo-relatives are probably the most selfied wild animals on Earth, because they do not fear people and appear to be constantly smiling.

Another Western Australian macropod (hopping marsupial) is the banded hare-wallaby. It is small, short-bodied and squat, with a square, short face, large round eyes and unusually marked fur: grizzled grey with a series of short horizontal bars over its back and rump and white eyebrow and moustache markings. Overall, it is very attractive. This diminutive species was once thought to be the last surviving member of the short-faced kangaroos – which included the largest known roo, the extinct giant *Procoptodon*, which reached 3 metres (10 feet) tall; however, recent studies have found that this is not the case. Instead, it belongs to a separate, ancient group of its own. Banded hare-wallabies were once found across large parts of southern and western Australia, but by the early twentieth century they had become extinct on the mainland and restricted to a couple of islands in Shark Bay. When I found myself surveying them there, I couldn't help but think I was following in the footsteps of William Dampier, the first English person known to encounter Australia's mammals* – dugongs on his first visit as a buccaneer in 1688, then banded hare-wallabies when commanding a naval vessel in 1699 (he also recorded seeing the tracks 'of a beast as big as a great mastiff-dog' in 1688,[24] presumably dingo footprints, which had also been mentioned by Dutch sailors in 1623). It is ironic that the first place that a European found banded hare-wallabies is also the last place that they were able to survive, following the environmental devastation caused by subsequent European invasion of Australia. Dampier wrote about the species he met in Shark Bay in 1699:

> The Land-Animals that we saw here were only a Sort of Raccoons, different from those of the West-Indies, chiefly as to their Legs; for these have very short Fore-Legs; but go jumping upon them as the others do (and like them are very good meat)'.[25]

Because of their markings and size, I think this is a reasonable comparison. We can also note that a common theme of sailors' accounts of animals they meet is that they prioritise information about their edibility. Understandably after months or years at sea, having the chance to eat fresh meat is high on the agenda.

So far, we have written descriptions of varying quality from the 1600s introducing dusky pademelons, tammar wallabies,

---

* But not the first English person on Australian soil. In 1622, a ship owned by the British East India Company called the *Tryall* went well off course trying to reach Batavia from Cape Town and wrecked on a shoal off the Western Australian Pilbara Coast (now called Tryal Rocks). More than thirty of the survivors stayed on nearby islands for a week before heading to Java in one of the ship's longboats. This represents the first time Europeans stayed in Australia for a protracted visit.

quokkas and banded hare-wallabies. However, it's fair to say that there is no evidence that any of these accounts influenced each other, or later encounters with marsupials. Perhaps this is because of the inherent difficulty of describing a completely unfamiliar animal to those who have not seen it – perhaps illustration would help bring them to life in the Western psyche. As they say, a picture paints a thousand words.

The first confirmed European depiction of an Australasian marsupial came in 1711, when the Dutch artist Cornelis de Bruijn met and drew a captive dusky pademelon in a Javan menagerie, having been brought from the Aru Islands, south of New Guinea. It's a very good likeness and it even shows the pouch.

So, by the time James Cook and Joseph Banks, aboard HMS *Endeavour*, came across kangaroos in north Queensland in 1770, one might expect that they would be able to relate their own observations to the information that had come out of previous voyages around New Holland and New Guinea. But they didn't. Even with de Bruijn's depiction, the confused descriptions of Australian fauna that had come before and the constant comparisons to cats, dogs, rats and raccoons inevitably led to their discoveries being more or less ignored. It isn't particularly surprising – the absolute unfamiliarity of a hopping marsupial to a seventeenth-century explorer meant that the animals were clearly beyond their descriptive ability.

Cook did not discover Australia, marsupials or macropods, but his voyage would do more to introduce them to the wider world than any had before.

*

In 2015 I curated an exhibition at the Grant Museum of Zoology called *Strange Creatures: The Art of Unknown Animals*, which explored the ways in which newly discovered exotic animals were described for people 'back home' in Europe. It centred around the first Western painting of an

Australian animal, *The Kongouro from New Holland*, painted by George Stubbs in 1772, on loan from the National Maritime Museum in London. On his return from the voyage on the *Endeavour*, Banks had enlisted Stubbs, arguably the greatest animal painter of all time, to paint the kangaroo (and also a dingo). He provided him with descriptions of the animal, and probably the pencil sketches made by one of the ship's draftsmen, Sydney Parkinson, who had died on the journey home. Not only that, but Banks had an 'inflated skin' and skull of a kangaroo, which have sadly not survived to the present day.* Borrowing Stubbs' painting for *Strange Creatures* has been one of the greatest privileges, and also the scariest responsibility, of my career. As I said in the exhibition:

> This painting helped begin Europe's relationship with Australian wildlife. Commissioned by legendary naturalist Joseph Banks, painted by celebrated animal artist George Stubbs, and based on findings from Captain Cook's famous voyage, this kangaroo truly captured the country's imagination.
>
> Stubbs' image became the archetype for representations of kangaroos for decades – reproduced and refigured prolifically. It may not be anatomically perfect, but this is how Britain came to know the kangaroo.
>
> It is an emblem of the age of exploration at the historical threshold of the European occupation of Australia. Nothing was ever the same again.[26]

This painting gets a lot of stick for the animal's odd pose and narrow tail, but given that he never saw a kangaroo for himself, I think Stubbs did incredibly well. In any case, it

---

* The subsequent history of the skin is not known. It's safe to say that the preservation techniques of the time where not very sophisticated. But the skull remained at the Royal College of Surgeons of England until World War II, when the college's building was bombed. A photograph of it survives.

was this rendering of the species that became the 'archetype of kangarooness', copied by many artists thereafter.[27] This painting influenced the European public's imagining of kangaroos for decades. It became so important that when it was put up for sale by its private owners in 2013, there was a diplomatic stand-off as the UK government blocked its sale to the National Gallery of Australia (allowing London's National Maritime Museum to buy it instead). Both governments insisted that it was a significant part of their country's heritage – which was obviously true – but as the artwork was painted in London by a British artist, commissioned by a British patron and had never left the country, the UK thought it was justified to stop it leaving. Nevertheless, Australia insisted that it was of more importance to the development of Australian imagery than to British history. The Australian press covered the story as an act of modern-day colonialism.[28]

This painting, and the circumstances of the kangaroo's description by Banks, would never have existed if it weren't for an extraordinary bit of bad luck in a very dangerous situation. The main published purpose of the *Endeavour*'s voyage was to take measurements of the transit of Venus across the face of the sun, which was to be observed from Tahiti. This would allow accurate calculations about the Earth's movement, which would prove valuable for navigation. However, there was a second, secret agenda – to confirm the existence of and to chart the elusive great southern continent, *Terra Australis Incognita* ('unknown southern land'). It was assumed that a massive expanse of land must exist in the southern hemisphere in order to 'counterbalance' the mass of land in the northern hemisphere. When the Dutch first reached the west coast of New Holland in 1616, they believed it was possible that it was the western extent of this great theoretical landmass. However, in 1642, when Abel Tasman sailed from Tasmania to the west coast of New Zealand, north to Fiji (he was the first European to reach all three places) and then around New Guinea to Java, his giant loop demonstrated that there was clear ocean south and east of New Holland. This meant that New Holland was nowhere big enough to counterbalance North America, Europe and Asia (however, although the fact that he could sail north from western New Zealand meant that the land wasn't continuous with New Holland, the latter's eastern extent remained unknown to Europeans). After Tasman, the only unexplored place that the 'great southern continent' could still be was between New Zealand and South America,[29] or far to the south.

James Cook established just how big New Holland was when he first caught sight of the southeast corner of the continent on 20 April 1770.* He sailed north, unable to find anywhere safe to land for nine days. When they did, Joseph Banks spent two weeks collecting plants in the area – which they would name Botany Bay – but sailing north again along the east coast they made only two other brief trips on land over the subsequent few weeks (for water, unsuccessfully). By this time, they had only recorded seeing one Australian mammal (probably a bandicoot) – the voyage was very nearly a zoological disappointment.[30]

Despite the lack of water, they had been incredibly lucky, having unknowingly sailed over the treacherous shoals of the Great Barrier Reef for weeks. That luck would run out on the night of 11 June 1770, as they ran aground on the sharp coral offshore from a site the crew named Cape Tribulation (which still bears that name, and is an excellent spot for seeing giant white-tailed rats), tearing holes in the *Endeavour*'s hull that could well have stranded them there.

After six days of pumping water and shedding weight, the *Endeavour* incredibly found a safe place to careen in an estuary

---

* He called it Point Hicks, following his practice of naming features after the man who first spotted the land (Zachary Hicks called 'land ho!' on this momentous occasion).

a long way to the north, where it stayed for six weeks to make repairs. If it weren't for this misfortune, the crew would never have seen any kangaroos and Banks wouldn't have instructed Stubbs to make the painting. Europe's knowledge of the animal would have been delayed until the First Fleet arrived in 1788 (initially in Botany Bay, on Banks' recommendation, although they were less taken with it than Banks and Cook were, and opted to make their settlement slightly further around the coast, at Port Jackson).

The encounters Cook and Banks did have were even less likely, given that the location they ended up in (the site of modern-day Cooktown) is more or less the northern limit of kangaroos' range.

Their observations of the animals during those six weeks follow the pattern of the previous explorers' marsupial descriptions, by including inappropriate comparisons to familiar species. Here follow some entries from Cook's own diary:[31]

June 22nd, 1770

Some of the people . . . had seen an animal as large as a greyhound, of a slender make, a mouse colour, and extremely swift.

It strikes me that although they had obviously seen it move, there is no mention of its most noticeable trait – that it jumps rather than runs.

June 23rd

This day almost everybody has seen the animal.

June 24th

I saw myself one of the animals which had so often been described. It was of a light mouse colour, and in size and shape very much resembled a greyhound; it had a long tail also, which it carried like a greyhound; and I should have taken it for a wild dog, if instead of running, it had not leapt like a hare or a deer. Its legs were said to be very slender, and the print of its foot to be like that of a goat.

Personally, I don't find any of the suggestions that kangaroos are greyhound-like very convincing. It's worth noting that there were two greyhounds on board the *Endeavour*, making these comparisons particularly unimaginative.

> July 8th
>
> Some of our men saw four animals of the same kind, two of which Mr. Banks' greyhound fairly chased . . . These animals were observed not to run upon four legs, but to bound or hop forward on two.

It's obviously difficult to disengage my modern experiences of kangaroos, but I find it hard to accept that the first thing anyone wrote about a kangaroo wasn't that it was a large animal jumping on just two feet – an observation it took them two weeks to make. Instead, it continued to be compared to an animal it resembles in very few ways – a greyhound.

> July 14th
>
> Mr. Gore . . . had the good fortune to kill one of the animals which had been so much the subject of our speculation. In form it is most like the jerboa [hopping desert rodents], which it also resembles in its motion, but it greatly differs in size, the jerboa not being larger than a common rat, and this animal, when full grown, being as big as a sheep.

Then follows a reasonable description of its head, legs and tail, and, as ever, how it tasted.

Despite the fact that Banks said, 'To compare it to any European animal would be impossible as it has not the least resemblance of any one I have seen',[32] these brief excerpts compare kangaroos to a greyhound (three times, in shape, size and tail), a wild dog, a hare and a deer (in movement), a goat (in footprint), a mouse (in colour), a jerboa (in movement and shape) and a sheep (in size). I'm surprised its taste didn't suffer a comparison (slightly like lamb's liver, if you ask me).[33]

Banks' own account, written shortly after they left Australia, deserves a lengthy quote as it contains a number of useful contributions, including the linking of possums ('phalangers') with animals already reported from New Guinea and the opossums of the Americas; a mention of dingoes ('wolves'); the source of the kangaroo's English name as a word from the Guugu Yimithirr language of that region (and that he definitely didn't recognise them as related to animals from previous accounts); that 'quoll' also comes from an Indigenous word; and that even he found his contemporaries' animal descriptions to be difficult to interpret.

> Quadrupeds we saw but few and were able to catch few of them that we did see. The largest was calld by the natives Kangooroo. It is different from any European and indeed any animal I have heard or read of except the Gerbua of Egypt . . . Another was calld by the natives Je-Quoll: it is about the size and something like a polecat, of a light brown spotted with white on the back and white under the belly. The third was of the Opossum kind and much resembling that calld by De Buffon Phalanger. Of these two last I took only one individual of each. Batts here were many. . . . Besides these Wolves were I beleive seen by several of our people and some other animals describd, but from the unintelligible stile of the describers I could not even determine whether they were such as I myself had seen or of different kinds. Of these describtions I shall insert one as it is not unentertaining. A Seaman who had been out on duty on his return declard that he had seen an animal about the size of

and much like a one gallon cagg; it was, says he, as black as the Devil and had wings, indeed I took it for the Devil or I might easily have catchd it for it crauld very slowly through the grass. After taking some pains I found out that the animal he had seen was no other than the Large Bat.[34]

Cook and Banks had found the east of the country to be far more inviting than the Dutch had on the west, south and north, and returned to England and made their reports. On the strength of these, the government sent a fleet to establish a penal colony in New South Wales in 1788, changing the country forever. In 1793, Watkin Tench, diarist of the founding settlement, wrote:

We have killed she-kangaroos whose pouches contained young ones completely covered in fur and of fifteen pounds weight, which had ceased to suck and were afterwards reared by us . . . It is born blind, totally bald, the orifice of the ear closed and only just the centre of the mouth open . . . At its birth . . . the kangaroo is not so large as a half-grown mouse. I brought some with me to England even less, which I took from the pouches of the old ones. This phenomenon is so striking and so contrary to the general laws of nature, that an opinion has been started that animal is brought forth not by the pudenda [genitalia], but descends from the belly into the pouch by one of the teats.[35]

This is from *A Complete Account of the Settlement at Port Jackson*. For some time, it was probably the most widely read description of colonial life in Australia. Likely influenced by the translation I mentioned of Francisco Pelsaert's report, as Tench says, finding such tiny babies in the pouch led the settlers to believe that kangaroos gave birth directly through their nipples. The phrase he uses to describe kangaroos to the people back in England is, 'contrary to the general laws of nature'.

You can't get much more othering than that. Like Pinzón's 'Monster' opossum, it put marsupials well and truly beyond the realms of normality, and into the alien.

*

It's not just historic opinions of Australian wildlife that paint an unreasonable picture. Today, it's incredibly common to hear it said that everything in Australia is dangerous. Sure, it has its complement of venomous organisms, including snakes, spiders, jellyfish, octopuses, ants, centipedes, stonefish, sting-rays and even trees (not to mention male platypuses), plus sizeable crocodiles and sharks. But so does everywhere else outside of Europe – and 'everywhere else' adds to that list several large land predators, from big cats to bears, and massive herbivores it would be perilous to approach. From this point of view, Australia is *less* dangerous than nearly every other continent. Even its native bees have no stings (but predictably, European honeybees have been introduced and gone feral).

I'll admit to having been briefly worried when I awoke in my sleeping bag out in the open in Central Australia and opened my eyes to see a dingo – Australia's largest remaining mammalian predator – looking down at me a couple of inches from my face. But that was nothing compared to the utter horror I felt when being momentarily charged by an American grizzly bear, which are twenty times bigger than a dingo. Or coming across fresh polar bear tracks in the snow when I was outside at night in the Canadian Arctic. Or having a lion's roar blast from a thicket I was standing next to when I was walking in Gujarat. Or even being in a tent surrounded by a mass of Italian wild boar. The fact is that Australia has only some of the types of animals that we might be fearful of, whereas most other parts of the world have many more.

Indeed, a recent report from the Commonwealth Scientific and Industrial Research Organisation (CSIRO, Australia's governmental science body) outlined how it is a myth that Australia is home to the world's most dangerous snakes: there is a tiny number of human deaths from snake bites there each year, but tens of thousands across Asia, Africa and South America.[36] I don't want to undermine the deadliness of some of Australia's snakes, either in terms of the power of their venom or their aggression – much of the difference in fatalities is due to the likeliness of people encountering snakes and the availability of medical interventions. Nonetheless, I find the unique synonymy of Australia with killer creatures noteworthy.

This commonplace attitude, to me, is just another form of colonial denigration; another unsubtle hint that Australia is uncivilised and primitive. People know that there are famously dangerous snakes in other parts of the world, but fail to mentally check whether the old trope about Australia being full of deadly animals actually stands up to scrutiny.

These subconscious biases have effects beyond light-hearted jibes suggesting that every living thing in Australia is trying to kill you. I mentioned in the introduction that inaccurate assumptions about marsupial brain size had unfairly written them off as inferior. Just as pernicious in slowing our under-standing of the evolution of life on Earth were the attitudes towards modern-day discoveries about the origins of most birds.

In his groundbreaking 2016 book, *Where Song Began*, Australian author Tim Low outlines the importance of Australia's role in our experiences of the natural world. Step outside or look out your window in most parts of the world and the vertebrates you are most likely to see are songbirds – the group that makes up around half of all bird species, including sparrows, finches, honeyeaters, wrens, tits or chick-adees, warblers, thrushes, robins, starlings, magpies and crows. Not only are the songbirds incredibly diverse, but they are often extremely numerous in their environments.

Australia is now known to be the place that this group first evolved: most of the world's birds have Australian ancestors. Not only that, but two other groups of extraordinary global importance also originated there: the pigeons and parrots. Thank you, Australia.*

When the Australian origin of these birds was first uncovered in the 1980s and 1990s, the discovery was largely ignored or rebuffed. The key reason for this appears to have been a reluctance to accept that Australia – rather than the northern hemisphere – could have played such an important role in evolutionary history. It didn't fit the widespread assumption that Australia was a world apart, with its primitive, evolutionary dead-ends going about their business in isolation from the rest of the world. Scientists were unwilling to accept that the opposite was true – the birds that evolved in and dispersed out of Australia changed the world. I've been on fieldwork with Tim Low and it is wonderful to see how his mind works – constantly unpicking assumptions and making connections. As he put it, the previously prevalent belief was that Australia's birds had evolved in the northern hemisphere and arrived in waves from Asia: 'This was a version of *terra nullius*, an empty land filling with good things from the north.'[37]

As I've already hinted at and will discuss further, I think the West's attitudes towards Australian wildlife have even closer links to the racist notion of *terra nullius* than this. Together, I see these unscientific biases towards the animals, as well as the othering linguistic tropes used to describe them, as part of a subconscious colonial mindset. As later chapters will demonstrate, this is having a very real and very negative impact on Australia today. But first, I'd like to spend some time celebrating the marsupials – the animals for which Australia is most famous.

---

* And then, a study in 2022 found that the ancestors of marsupials and placentals originated in the Southern Hemisphere, with key fossils from Australia. They later dispersed to the Northern Hemisphere where the two groups then emerged. This means that we should also be grateful to Australia for all the world's mammals. (Flannery et al, 2022b)

7

Marvelling at Marsupials

Despite popular conception, marsupials are not uniquely Australian. Nearly a third of all marsupial species today – around 100 – live in the Americas. Nearly all of these are called 'something opossum'. The term 'opossum' derives from the Algonquian word *apossom*, first recorded in 1607 by Captain John Smith, of Pocahontas fame. Writer William Strachey's report of his travels to the Virginia colony in 1612 included a dictionary of the local language, which defined the opossum: 'Aposon, a beast in bignes like a pig and in taste alike'.[1] While this demonstrates that the Virginia opossum was a food source for early colonists, it also suggests that their desire for food left them prone to seeing things. Opossums are not like a pig in bigness: a large opossum weighs 5 kilograms (11 pounds), a fraction of the weight of a pig. In fact, Smith's original description was more accurate, and Strachey's line appears to be a poor replica of it: 'An opossum hath an head like a Swine and a tail like a Rat, and is the bigness of a Cat. Under her belly she hath a bagge, wherein she lodgeth, carrieth, and sucketh her young'.[2]

In truth, opossums range in size from the shrew opossums and mouse opossums (which are – ahem – shrew-sized and mouse-sized) to the large American opossums (which are cat-sized). There are also four-eyed opossums, woolly opossums, short-tailed opossums and slender opossums. One can't help but feel that no one has ever put much effort into thinking up the names of these animals.

The scientific name for the main group of opossums, Didelphidae, references their pouch – it means 'two wombs', referring not to the fact that they have two uteri like Australian marsupials (which they do), but that they have a 'second womb' on the outside, in the form of a pouch.

Outside their native range, few people know that they exist, with the possible exception of the Virginia opossum in the USA – although I wonder how many people know that Virginia opossums are marsupials. This species is commonly considered a trash animal with no popular appeal (it is also regularly used as a prop in derisory depictions of backcountry 'hillbillies', so it seems that American marsupials can be involved in hierarchical social stereotypes just like Australian ones). 'Playing possum', or pretending to sleep, originates with them, as they feign death as a defence mechanism to put off attackers. The (distantly related) Australian possums are named after the American opossums.

Why are these American marsupials largely ignored by the world at large? Conversely, it is often assumed that all Australian mammals – except maybe the platypus and echidnas – are marsupials. In fact, marsupials only make up half of the Australian mammalian land fauna. A quarter is made up of native rodents and the final quarter consists of bats, both of which are placental mammals.

Nonetheless, marsupials have evolved to occupy every corner of Australia's vastness, and a massive range of ecological niches. There are (or were, before their recent extinctions) a full gamut of grazers, browsers, fruit eaters, nectar drinkers, truffle specialists, insectivores, scavengers and predators large and small. Some can sprint, some can hop, some can dig, some can swim through sand, some can climb and some can glide. They thrive on cliffs, on temperate grasslands, among boulders, in rainforests (cool and tropical), in deserts, in monsoonal woodlands and savannahs, in marshland and on mountains. What they don't do, however, is fly, and while many marsupials are capable swimmers – including my favourites, the wombats – no fully aquatic marsupials have evolved (platypus-pioneer David Fleay suggested that one possible reason for this in Australia is that the platypus occupies the aquatic niche so well, that marsupials haven't had a chance in this habitat*)[3].

*

While marsupials have never evolved 'true' flapping flight, the ability to glide has evolved multiple times among the possums, as it has among rodents elsewhere in the world. By stretching a membrane between their limbs and body, these animals have independently produced winglike structures in various different groups.

Greater gliders are possums the size of a cat, although much lighter, which can glide for an astonishing 100 metres (330 feet) between trees, without a single 'wingbeat'. Their massive, fluffy tails – longer than their bodies – provide a level of stability and steering that enables them to turn a full 90 degrees mid-glide.

Gliding evolved separately in the group that contains the squirrel-sized sugar gliders and slightly larger yellow-bellied gliders, which become flat squares when they launch into the air, as they have full wing membranes (called patagia) extending from their ankles right up to their fingertips. These membranes can carry a yellow-bellied glider for 140 metres (460 feet) as they travel between trees that they cut little Vs into for feeding on the sap. By contrast, greater gliders' wing membranes end

---

*The yapok, or water opossum, of South and Central America is a semi-aquatic marsupial, with webbed feet and a pouch that can be closed up tightly. These opossums also have a unique hunting method – swimming with their hands held out in front of them, they feel for fish and invertebrates to grab. Their fingertips have little fingerlike projections of their own, to enhance their sense of touch.

An 1865 illustration of a sugar glider.[4]

at their elbows, so their 'parachutes' are proportionately smaller. When they are airborne, greater gliders fold their hands under their chins, elbows out, in what has to be the most comical pose of the world's 'winged' mammals.

The feather-tailed gliders, in yet another group, do it differently again. These tiny marsupials are only about 5 centimetres (2 inches) long (they are the smallest of the world's gliding mammals) but can glide for over 25 metres (80 feet). Their tails are almost naked on the top and bottom but have long rows of thick hairs sticking out on either side, giving them a close resemblance to feathers. Their toes have special ridges and sweat glands to help them grip and stick to smooth tree bark.

The various groups of possums are by no means entirely composed of gliders. Most of the approximately thirty Australian species spend much of their life in trees and have notable adaptations for climbing. For example, they have opposable big toes for grasping branches (which among mammals is only otherwise seen in primates and opossums), and most possums – as well as their distant relatives, koalas – can do something similar with their hands, albeit with two fingers on one side of the branch and three on the other.

Curiously, the wyulda has the same hand arrangement as primates, with just an opposable thumb. Some cuscuses – woolly-coated possums from the tropics (more than twenty species live in New Guinea, Indonesia and surrounding islands, while two live in Far North Queensland) – resemble monkeys even more closely, with short, pink, naked noses and small ears. Indeed, historic accounts of 'monkeys' from this region surely refer to these species.

Most possums also have a prehensile tail, which allows them to grip branches like a fifth limb. The tail can have fingerlike ridges for extra security, and in many species it can support their entire weight when they dangle from it.

The majority of possums are herbivores, but unlike most plant-eaters, which live on the ground, their eyes face forwards on the front of their heads. With eyes on the side of their heads, animals like deer, horses, goats, antelope and kangaroos have a very wide field of vision for spotting predators coming at them from almost any angle – but in order to judge distance best, you need your eyes facing forwards. In this way, with binocular vision, possums can accurately jump between branches tens of metres above the ground and reliably land in the correct spot. In their evolution, this has obviously been found to be more important for living up in the trees than watching for predators, although they have many, from quolls to pythons and from monitor lizards to owls.

In most parts of Australia, you go spotlighting and see eyeshine from an animal in a tree, and as soon as you've judged what size the animal is, there is normally only a small handful of species that it might be. In Tasmania, for example, unless you're lucky enough to find a quoll up a tree (which doesn't happen often), you're only likely to be looking at a sugar glider (but according to a 2021 study, this population may actually become known as Krefft's gliders)[5], a common brush-tailed possum, an eastern ring-tailed possum or a nocturnal bird. However, wandering under the giant trees of tropical

Queensland, such as in the Atherton or Mount Carbine Tablelands, you have more than ten possum possibilities climbing among the branches. In addition to those, there are also giant tree-rats, tree-kangaroos, quolls, tree mice, several fruit bat species, geckos, 5-metre (16-foot) long pythons *and* owls and frogmouths (not to mention what might be going on down on the ground). They are not all found in all habitats in these areas, but this is an exceptionally diverse spot for climbing mammals, particularly ring-tailed possums, which are the greater glider's closest relatives. In this small corner of the wet tropics, you can find green ring-tailed possums, Daintree River ring-tailed possums, eastern ring-tailed possums, lemuroid ring-tailed possums (which closely resemble lemurs) and Herbert River ring-tailed possums, plus the gliders. And that's just the ring-tailed possums. Walking around there with a head torch at night is almost overwhelming for mammal nerds, so much so that you have to remember to go to bed at some point. This region has some particularly nasty hazards at ground level, too, which you must remind yourself to look out for while also scanning the treetops. Wait-a-while vines, for example, have hooked spines covering their thin stems, which if you brush past snag into your skin and clothes. If you try and brush it off, or walk on through it, it ends up ripping through you in a really unpleasant way – as if Velcro was made from thousands of tiny blades. If you get snared, you have to stop immediately (and 'wait a while') in order to extricate yourself without moving the vine in a way that increases its contact with you. Stinging trees are even more pernicious. Their leaves resemble harmless raspberry bushes, but if you come into contact with them, they are covered in millions of fine silica-tipped needlelike hairs, which inject venom. The ensuing sting is excruciatingly painful and can reoccur for months afterwards. This is not to mention the green ant nests that dangle from low branches and somehow manage to completely cover you with ants with just the slightest of touches. They then proceed to bite, particularly in hard-to-reach places like in amongst your hair (curiously, if you crush them while trying to pick them off, they give off a very pleasant lime scent which in small quantities is even pleasant to taste).

It is challenging for small mammals to live entirely off leaves, as they normally need to be eaten in very large quantities to give the animal enough energy, and smaller animals typically have higher metabolisms than larger animals. Nonetheless, at least some species of ring-tailed possum are able to get around this problem. Leaves, particularly tough ones like eucalypts, are extremely difficult to break down. In order to do this, ring-tailed possums have very effective teeth for pulverising the leaves into tiny fragments, and chambers in their guts that house symbiotic bacteria which break down the tough fibres. However, these chambers are positioned after the point in the intestines where most of the nutrients are absorbed into the blood. This means that by the time the complex molecules have been broken down into chemicals that the possum can 'use', they are too far along their digestive system to do anything with them. And so the ring-tailed possums do the same thing that rabbits do (which have the same problem) – they eat their poo.[6] During the day, while they are resting in the nest, they produce a soft poo that contains all the useful nutrients that their gut bacteria have prepared, and by eating those poos the possums give them another run through the parts of the guts that can absorb them. Then, while they are out foraging at night, the animals produce a different, hard poo that is just waste products.

One species that spends less time in the trees is the rock ring-tailed possum, from tropical areas of the Northern Territory and the Kimberley. These possums have independently evolved to live a similar life among the rocks as the scaly-tailed possum, to which it is not closely related, but they look very much alike. They are highly social animals, living in communicative little groups. When they are feeding, the adults take turns standing guard as a sentinel while the rest of the group forages. If they spot any signs of danger, they will make calls and bang their tails against a branch to raise the alarm. Meerkats have become famous for very similar behaviours – maybe it's time rock ring-tailed possums got a share of the limelight for this cooperative activity.

Male rock ring-tailed possums exhibit more paternal care than most marsupials do (or indeed many placentals). The babies are shepherded around, sandwiched closely between the mother and father. Up in the trees, the male and female help the youngsters cross gaps by forming a bridge out of their two bodies between the branches, so that the young can walk safely across their backs.[7] Presumably, these behaviours have evolved as both parents are needed to successfully raise the babies down on the ground where predators are more of a risk.

At around 4 kilograms (9 pounds), the cuscuses of the tropics and the similarly sized brush-tailed possums are the largest members of the possum group, but at the other end of the scale – at just 5–10 grams ($^1/_6$ –$^1/_3$ ounces) – we have the honey possum.

Honey possums (also known by one of their Indigenous names, noolbenger) are restricted to the far southwestern corner of Western Australia. With more than 5,500 flowering plant species that are found nowhere else on Earth, this is one of the world's thirty-six officially recognised biodiversity hotspots. One might imagine that this explains why the honey possum is found here: it is the only non-flying mammal on the planet known to rely entirely on nectar and pollen. But while it is true that it requires enough plant species to ensure that something is in flower on every day of the year, honey possums actually do almost all of their feeding on fewer than ten species.

These miniature marsupials have no close relatives and are unusual among climbing mammals, as instead of having gripping claws they have broad fingertips with nails, and an opposable big toe, rather like a primate. Indeed, their scientific name, *Tarsipes*, means 'tarsier foot', in reference to the agile little primates of Southeast Asia.

Being so small and nimble means that they can climb around flowers in order to feed. Among their favourites are the huge, nectar-laden cones of *Banksia* plants, named after our old purveyor of platypuses, Joseph Banks. *Banksia* cones, and many of their other food plants, are far larger than the possums themselves, and the marsupials play an important pollinating role as they crawl over them. *Banksias* can be so nectar-laden that it's possible for humans to lick a sizeable sugary serving from these glistening cones.

To feed, honey possums flick their tongues in and out of their extraordinarily long snouts two or three times per second. These tongues are around 2 centimetres (¾ inch) long, re-inforced with a stiff ridge underneath, and are guided by the possums' long, projecting lower incisors and little flanges on their lips. The tip of the tongue has a brushlike structure, which picks up the nectar, and the tongue's surface has little bumps to collect pollen. As it is drawn back into their mouths, the pollen is scraped off by ridges on their palate and their reduced canines. In this way they can consume almost their entire body weight in nectar each day, plus a sizeable portion of pollen.[8]

Not much larger than the honey possum are the pygmy-possums. Among them, the mountain pygmy-possum is particularly interesting. They were first described in the nineteenth century from fossils, and living animals were not found until 1966, when they turned up in a ski hut. The entire global population is now only known from three isolated locations covering a total of 6 square kilometres (just over 2 square miles). They are constantly under threat from a bad bushfire season. Living in the mountains of Victoria, where snow can cover their habitat for half the year, they are the only marsupial known to hibernate for long periods – up to seven months of the year, during which time their body temperature can drop to just 2 °C (35 °F) for up to three weeks at a time (they then warm up for a day before dropping back down).

This species primarily lives on the ground among boulders and is the only Australian mammal restricted to alpine habitats. Here, they mainly eat insects, and are particularly partial to the migratory bogong moths that come to the region every year. Rich in fats and protein, these are the perfect meal for a small animal to feast upon when it is trying to fatten up, and mountain pygmy-possums can double their weight by gorging on the moths.

I mentioned before that it can be disappointing when a species you are not targeting turns up in your traps. It is very hard to feel any such emotion when a pygmy-possum does this. On one trip, I was helping Rosie Hohnen (now Dr Rosie, having successfully completed her PhD involving the scaly-tailed possums) to locate sites where the critically endangered Kangaroo Island dunnarts might still cling on. Dunnarts are a group of miniature marsupial carnivores – about 10 centimetres (4 inches) long – and this one was only known from a handful of spots on Kangaroo Island off the South Australian coast.* Rosie was working to add to that list of locations, so we set remote cameras and dug a series of pitfall traps in likely-looking habitats and hoped for the best.

Pitfall traps are a very simple means of locating small animals, particularly small mammals, lizards, frogs and invertebrates. First, you dig two pits in the ground to each

---

*Tragically, all of those locations were affected by the catastrophic 2019/2020 bushfire season; however, the team at the Kangaroo Island Land for Wildlife biodiversity programme were able to mobilise quickly after the fires passed through and found a new location. With help from the Australian Wildlife Conservancy, they rapidly fenced this area to protect the survivors from feral cats, which move in after fires to sweep up any fleeing wildlife. Since the fires, intense survey work by the South Australian government has managed to detect more dunnarts at a good number of other sites within the fire scar – not only in small unburned patches, but also in sites that had been completely burned – and remove many cats from these spots. So that's some happy news.

accommodate a bucket or a closed-ended tube deep enough that your target species can't jump out, then you place the buckets into the pits with their tops level with the ground. Next you erect a fence between the two pits, so that it runs straight across the tops of the buckets – we typically use rolls of the tough, flexible plastic used in damp coursing. Finally, you put a little soil and some stones and leaves into the buckets to provide anything you catch with some shelter. The idea is that when an animal is walking about minding its own business, it will come across the fence. Assuming you haven't left any tiny gaps underneath, it will then follow the fence and with any luck will drop into one of your buckets at the end. Then, you simply keep checking very regularly while the traps are open, and see what has stumbled in. After taking whatever measurements you need, you release them.

Sadly, we didn't catch any Kangaroo Island dunnarts while I was with Rosie, but we did find a number of western pygmy-possums. Looking for critically endangered animals can sometimes be depressing work, as by its nature it usually has a low success rate, and hours of digging in what can be rock-solid earth often comes to nothing (although negative results are useful, if disheartening). However, the appearance of these tiny possums – which feed on insects, pollen and nectar – is enough to rouse anyone's spirits. They look like they have been designed by a cartoonist whose brief was to come up with the most adorable creature that their imaginations could muster. Although only mountain pygmy-possums hibernate, any of the five species can enter daily torpor if it gets chilly. This is when they drop their core body temperatures down to a little over 10 °C (50 °F), lower their metabolism to save energy and curl up into a little ball. They are tiny – just 6 centimetres (2 inches) long, and oblong, with big round eyes, a petite snout and huge ears, which they fold down over their face while sleeping. They also have long tails, which they spiral into a perfect coil, like the neatest rope on a ship's deck. When you find them in torpor, they sit curled up in the cup of your palm, slowly rousing from your body heat. The heart melts.

*

Those Kangaroo Island dunnarts are just one of over seventy carnivorous species from Australia and New Guinea that make up a major division of marsupials called dasyurids. Despite the fact that there are so many of them – comprising roughly a quarter of Australia's land mammal species (excluding bats) – and that they are found across all Australian habitats, sometimes in great number, they do not enjoy a very public profile – aside perhaps from the Tasmanian devil. I suspect few people have heard of planigales, mulgaras, dibblers, dunnarts, ningauis, antechinuses, pseudantechinuses, phascogales, kowaris, kultarrs, kalutas or quolls.

With most individuals weighing in at 7–9 kilograms (15–20 pounds), devils are by far the largest of the dasyurids. A real beefcake male might approach 14 kilograms (31 pounds), but they are unusual. By contrast, most of the other dasyurids weigh less than 100 grams (3½ ounces). These are often called 'marsupial mice', but they have little in common with mice beyond their size. 'Marsupial shrew' would be more fitting (but you know how I feel about this kind of comparative naming), as both groups are carnivores. Most dasyurids, particularly the smaller ones, have sharply pointed, almost conical snouts and very large ears, often shaped like human ears.

The very smallest is the long-tailed planigale, which can weigh less than 3 grams (1/10 ounce). A large devil is nearly 5,400 times heavier than a large long-tailed planigale.* Three grams is near the lower limit for how small an endothermic ('warm-blooded') animal can be, because of the ratio of their

---

* When we look at the smallest of all mammals compared to the largest, this kind of comparison becomes even more amazing: 150 million Savi's pygmy shrews weigh the same as a single blue whale. It's one of my favourite mammal facts.

surface area to body size (small objects have a relatively large surface compared to their volume). Below that weight, mammals are likely to lose more heat through their skin than they can generate internally. They are so light that I have come across tiny planigales in pitfall traps that almost escaped because they had climbed up a single grass stem that had fallen into the pit.

Planigales are adapted to a life in tiny spaces. Long-tailed planigales are found on the black-soil floodplains of northern Australia. In the wet season, these plains may be completely underwater (and the planigales have to move to higher ground), but in the dry season the soil cracks to extraordinary depths – reaching 2 metres (6½ feet) below the surface. A planigale's head and body are flattened to live in these little crevices – their skull is only 4 millimetres (¹/₆ inch) deep. The jaw is lined with a row of tiny sharp teeth, and they are not afraid to use them.

Of all the animals I have worked with, by far the most ferocious are the planigales. No species has attempted to stare me down, puff itself up and go on the attack with the savagery of these micro-marsupials. Their teeth are around a millimetre long, so there is no risk of them breaking human skin, but you cannot fail to admire their chutzpah. Although they can't overpower an ecologist, this boldness presumably comes in handy when they are on the hunt.

Like many small dasyurids – who are all similarly plucky – planigales hunt invertebrates and even small mammals and lizards that are larger than them. This includes species that would have more than enough venom to kill such a tiny mammal, including centipedes, scorpions and spiders. Some species are so small that if the planigale fails to give a killer bite to the head and neck when it attacks, it can find itself carried off by its prey, for instance being dragged along or pulled into the air by a hopping grasshopper.[9]

Ironically, devils, despite their reputation – which is in no small part thanks to the far-from-accurate portrayal of Taz in the *Looney Tunes* cartoons – are comparative darlings compared to their minuscule cousins. When being handled, few animals are more docile than devils. Since they have the strongest bite force of any living mammal, however, we don't take any chances.[10]

After the devil, the next largest dasyurids are the quolls, four species of which live in Australia. Following the relatively recent extinctions of devils and the 2-metre (6-foot) long and 17-kilogram (37-pound) thylacine from mainland Australia, the stunningly handsome spotted-tailed quoll is now the largest carnivorous marsupial on the mainland, at around 4 kilograms (9 pounds).* They are found in a band around the southeastern corner of mainland Australia and in a pocket of tropical north Queensland (this species lives in Tasmania, too, but thankfully the devil is not yet extinct there, although the thylacine is). Eastern quolls are mainly restricted to Tasmania, having probably gone extinct on the mainland,[11] although some have since been reintroduced. Western quolls once lived across 70 per cent of Australia, in the arid and semi-arid areas of the west and centre, but became restricted to the very far southwestern corner until a very recent reintroduction programme got underway in South Australia. Northern quolls are found in several spots across the north, although their range has also contracted dramatically. The trend for quolls declining since Europeans arrived is clear, and the major culprits are introduced cats and foxes.

There are also two species in New Guinea: the bronze quoll and New Guinea quoll, which are the largest indigenous carnivores remaining in New Guinea (aside from the New Guinea singing dog, which has an important role in the ecosystem there, but as it probably arrived by boat some people disagree as to whether it should be labelled as an indigenous species – the same is true of its relative in Australia, the dingo).

---

* With that said, a trial reintroduction of Tasmanian devils into a fenced reserve in New South Wales got underway in late 2020, potentially paving the way for the first wild and free devils on the mainland in 3,000 years.

A dark-coated eastern quoll (they can also have fawn-coloured fur), printed in 1827 from a drawing by the English-born artist John Lewin. Lewin is an important figure in the history of Australian art; however, this quoll has been given thinner and longer legs than they have in life.[12]

Although quolls are in the larger size-range for dasyurids (ranging between 300 grams and 4 kilograms, or 10½ ounces to 9 pounds), they more closely resemble the planigales than the devils in terms of feistiness. The bronze quoll, which was only described in 1987, was given the scientific name *Dasyurus spartacus* after the gladiator who rose up against the Romans, in reference to the quoll's ferocity.

Medium-sized mammals make up the largest element of a quoll's diet, and they can bring down wallabies, pademelons (forest-dwelling relatives of wallabies) and possums that are much larger than them.

Quolls have elongated bodies, elegantly pointed faces, relatively short legs and long, furry tails. They are among the few mammals that have coats heavily speckled with white spots against a dark background (how many more can you think of?*). On one field trip to the Kimberley in the dry season, Rosie Hohnen and I spent the hottest parts of the day, in

---

*Some species of deer, baby tapirs, some dolphins, some colugos, spotted and harlequin bats and a few rodents, including pacas and ground squirrels.

between our morning and evening trap-checking tasks, crouched around a computer in a sweaty Portakabin (which the Australians call a donga, particularly there in the northwest). We were studying camera-trap images of endangered northern quolls in order to determine whether individual animals could be reliably identified by their spot patterns. By drawing recognisable shapes around groups of spots on a quoll's coat – treating them like star constellations ('this one looks like a spaceship/a stegosaurus/Pac-Man') – and seeing if we could find the same animals over multiple nights on different cameras, we demonstrated that each quoll's spots are unique.[13]

If ecologists have the means to consistently identify an animal as an individual, it enables them to determine how large their home range is, and how many animals live in a given area. Traditionally, this requires resource-intensive catching, marking (with a microchip or ear tag, for example), releasing and then re-catching lots of animals over an extended period. By demonstrating that individuals can be recognised on camera without artificial marking, these important questions can be answered in a far less invasive way. Computer algorithms can now be trained to recognise quolls from their spots, so we don't have to do it with pencil and paper anymore.

Most dasyurids either follow the polecat-like model of a quoll, with a long body and short legs, that of the shrew-like planigales, or something in between with varying levels of rotundity. One significant departure from that general rule, however, is the kultarr. Found in small numbers across Australia's arid zone, kultarrs have a highly pointed snout, massive ears, large bulging eyes and an 8-centimetre (3-inch) body that roughly looks like other small dasyurids. But, this is all perched on extraordinarily long hindlegs. Most of their legs' length is provided by the main part of their feet, which are almost a third the length of their body. Their short toes are flat on the ground, but their feet are angled upwards. Their rear limbs are so long that when they were first described they were thought to hop on their toes and were accordingly called marsupial jerboas. However, it is now known that they move on all fours, bounding from their unusual hindlimbs onto their far shorter (but straighter) front legs. It is thought that this arrangement makes them extremely agile while avoiding being bitten by dangerous animals they themselves are trying to bite, such as centipedes and spiders.[14]

Kultarrs have a long, thin, tufted tail – presumably to act as a counterbalance to aid in their agility – but many dasyurids are characterised by having carrot-shaped tails. This is because many dasyurids store fat in the base of their tail. A well-fed animal can have an extraordinarily round tail, which tapers gradually down to the tip. Two species that take this to the extreme are the aptly named fat-tailed dunnart and fat-tailed pseudantechinus, which, if conditions are good, can be dragging a veritable sausage behind them.

For many dasyurids, the width of the base of their tails is a good indicator of their health. This is handy for ecologists because measuring the fatness of one animal's tail tells us about the health of that individual; and measuring many tails from a population gives us a good indication of the functioning of the entire ecosystem (how much food is available). For studies into the impacts of disease, like devil facial tumour disease, these measurements are vital.

Depending on the study, ecologists can take a lot of measurements from their study animals, and one common one is the width of the males' testes, as an indicator for their reproductive readiness and potentially their status in any social hierarchies (bigger balls means more testosterone*). The typical tool for measuring distances across three-dimensional objects is a set of callipers. Callipers are shaped like a capital F, where the top horizontal bar is fixed in position, and the lower bar

---

* For the record, it would be unwise to draw any conclusions for the human realm from such ecological studies.

slides up and down on the vertical bar. Measurements are taken by positioning the top bar on one side of an object and sliding the lower prong tightly against the other side. A scale on the vertical bar gives the measurement to a highly specific tenth of a millimetre (0.004 inches). Despite this degree of accuracy, there is a persistent suspicion among mammal ecologists around the world that different groups of ecologists consistently and predictably achieve different results when measuring the size of mammal testes. Specifically, male ecologists are believed to give larger readings than females measuring the same scrotum, because men are less inclined to squeeze the callipers tightly against the balls. It seems that no formal zoological study has tested these rumours (something I plan to rectify), but it has been done with medical practitioners measuring human manikins representing different stages of puberty with known testicular dimensions. It was found that the males in the study produced more accurate measurements of that particular part of the anatomy, but they weren't consistently larger or smaller than the female measurers' results.[15] Presumably this is down to the relative familiarity with the subject matter.

Of course, such measurements of genitals or tails can only be carried out if you temporarily catch the animal, and the easiest way of doing that for the small dasyurids is through pitfall trapping. However, many dasyurid species spend a lot of time up in the trees, including several species of antechinus. Antechinuses are generally ruddy or grey-brown, range in size from around 20 to 120 grams (¾–4 ounces) and are generally stockier than the planigales and dunnarts. Their bodies increase in girth back towards their rumps, so that the conical appearance of their noses can extend all the way down their bodies to the region of their hips.

Antechinuses and their close relatives the phascogales (which are shaped more like slender squirrels) are so at home in the trees that they can be seen running at speed along the underside of branches. Phascogales can rotate their ankles through a full 180 degrees, which allows them to run head first down tree trunks with their claws still hooking into the bark. This adaptation is shared with other expert climbing mammals, including chipmunks, margays (a kind of cat), kinkajous (related to raccoons), binturongs (a type of civet) and fishers (in the group with weasels and badgers). Despite being less than 20 centimetres (8 inches) long, phascogales can leap 2 metres (6½ feet) through the air from branch to branch. A huge 'bottlebrush' covers the last two-thirds of their tail, which may be useful as a counterbalance as they go about their treetop activities. Phascogales will feed on whatever they come across (apparently including live chickens, when they come down to the ground around human settlements), prising back bark with their teeth in order to hunt invertebrates hiding below, and taking small mammals, birds, lizards and frogs, as well as nectar from flowers.

Back down on the ground, the pseudantechinuses – which look a lot like antechinuses but are actually more closely related to devils and quolls (because evolution does that kind of thing) – are mostly found in rocky habitats and are also decent climbers. My favourite, the Ningbing pseudantechinus (named after the cattle station where the first museum specimen was found), lives among the red sandstone escarpments of the Kimberley and perhaps combines perfectly the typical dasyurid features: massive ears, a cone-shaped snout, a chubby (but rather long) carrot-shaped tail and decent climbing abilities.

*

If the dasyurids are some of the little-known species, without question the most famous marsupials are the kangaroos. Appearing on the national crest, the logo of one of Australia's largest global companies, Qantas, and surely every tourism advert the country has ever produced, kangaroos are true

Australian icons. Arguably, no nation is as firmly associated with an animal in people's minds as Australia is with the kangaroo.

After first having spent a lot of time in Tasmania – where even in the hills overlooking the centre of Hobart, the state capital, it is extremely easy to find hopping marsupials – when I later went to mainland Australia I was surprised and disappointed by how infrequently one comes across kangaroos, even in the wildest regions of the outback. Such is the strength of the kangaroo 'brand identity'.

Kangaroo is a term that typically refers to any of three species: the western grey kangaroo, the eastern grey kangaroo and the iconic red kangaroo, which at over 90 kilograms (198 pounds) for a big male, makes it the largest living marsupial. But these are not the only large hopping marsupials. The three species of wallaroo (the common wallaroo or euro, the antilopine wallaroo and the black wallaroo), which are found in rockier, hillier country than kangaroos, are actually the red kangaroo's closest relatives (closer than reds are to the two greys). This goes to show that just because certain species share the same general common names doesn't necessarily mean that they are closely related.

Kangaroos and wallaroos are the largest members of the major marsupial group known as macropods, which means 'big foot'. These are the hopping marsupials (although because nature never likes it when rules work cleanly, there is one pigeon-sized species – the smallest – called the musky rat-kangaroo, which bounds rather than hops). Found in Australia and New Guinea, macropods (of which there are more than 70 species) include the wallabies, rock-wallabies, tree-kangaroos, pademelons, quokkas, hare-wallabies, banded hare-wallabies, forest wallabies, nail-tailed wallabies, bettongs and potoroos, all of which are essentially variations on the general kangaroo theme, adapted to different lifestyles and habitats.

Again, except for the musky rat-kangaroo, which has a full

complement of five toes, macropods are distinguished by the unique layout of their feet. When describing animal digits, we start from the one nearest to the centre of the body and call that the first digit (which would be the thumb and big toe in humans) and move outwards to the fifth digit (our little finger or pinkie and little toe). In their evolution towards a hopping lifestyle, macropods have lost the first digit on their feet; the bones of their second and third digits are slender little rods bound up together so they look like one single toe but with two claws; their robust fourth digit is massively elongated and is positioned in a line with the rest of the leg and all the power is driven through it; and their fifth toe is much smaller. If you hold your hand in front of you, point only your index finger and place your thumb next to it, the shape made by those two digits is roughly the shape of a kangaroo's foot (only in mirror image).

That feature where their second and third toes are bound up together is called syndactyly. It is also found in macropods' close relatives, the possums, wombats and the koala, as well as in bandicoots. It is thought to have evolved independently in those two groupings – the reason why is not well understood, but it may result from the way the digits develop in the embryo, rather than there being a particular advantage in two toes functioning as one.[16]

Macropods are famous for the way they move. At slow speeds, most species do something called pentapedal walking, which means 'walking on five feet' – the fifth being their tail. They reach forward onto their short front legs, lift their bodies up by pushing down into the ground with their strong tails, and swing their big back legs forwards.

When moving fast, they hop on their powerful back legs. Their hops are an extremely efficient way of travelling – far more so than placental mammals that gallop. As kangaroos come to land at the end of a jump, their elastic tendons recapture 70 per cent of the energy that they spent on the initial leap, which can be 'reused' as they take off again in the next jump – like a coiled spring. A dog or monkey's tendons only capture about 25 per cent of that energy as they run, which is similar to springhares, the large hopping African rodents.[17]

Kangaroos switch from pentapedal walking to hopping at around 10 kilometres per hour (6 miles per hour), as this is the point at which hopping becomes more efficient than walking. For most animals, the faster they move, the more energy they expend. However, up until speeds of around 35 kilometres per hour (22 miles per hour), hopping kangaroos' energy costs remain the same. At that speed – the most efficient for them – they are using only half the amount of energy of other mammals of similar sizes. They move faster not by hopping more rapidly, but by hopping further, and a large kangaroo can leap 4 metres (13 feet) in a single bound. They can exceed speeds of 50 kilometres per hour (31 miles per hour).

The success of this group is presumably linked not only to the efficient way that they move, but also to the way they go about acquiring their supply of energy. Red and eastern grey kangaroos are grazers, eating grass like a lot of the farm animals that people eat. And like cattle and sheep, kangaroos have a compartmentalised stomach, with a chamber that contains microorganisms that break down the tough plant fibres found in grass. However, not only is the kangaroo system more efficient at extracting energy from grass than that of cattle and sheep, but it is also better for the planet. When ruminants like cattle and sheep digest plant fibres, they produce a massive amount of methane and burp it out, making one of the most significant contributions to the current climate crisis, through the greenhouse effect. By contrast, kangaroos produce very little methane. If the world could be convinced to replace cows and sheep with kangaroos as major meat-providers, carbon emissions would fall dramatically. And kangaroo meat is far leaner and healthier than beef and lamb.

Getting nutrients from plants is hard work, because they

require so much grinding to break down. Not only that, but grasses contain a lot of silica – which are essentially little granules of glass. This is tough going on herbivores' teeth, which get worn down throughout their lives. This is worse for animals that live in sandy places, as any grains that inevitably get into their mouth wear away their teeth surfaces even faster. As such, most grazers have tall ridges and cusps on their chewing teeth. And indeed, grazing kangaroo teeth are highly cusped, but they have another trick up their sleeve, too. Rather than having one set of baby teeth followed by one set of adult teeth, like in humans and most other mammals, kangaroo adult teeth appear in waves. While they still technically only have two 'sets' of teeth, their adult molars move horizontally forward along the jaw throughout their lives. A young adult will have two molars, later in life another two will move in from behind and they can have four at once, but later on the front ones will wear out and the nubs will be pushed out by the forward movement of the ones behind, leaving just one or two in old age. This maximises the period in which they have fresh, unworn molars, without increasing the overall number of teeth. Elephants do a similar thing: they only have one tooth in each side of the jaw at once. As it wears down, another one pushes in behind and replaces it. This happens six times in their life, and if they live long enough to grind down their sixth set, they then die of starvation.

Adaptations like that, along with experience of crippling toothache and knowledge of the extraordinary lengths people go to to replace their failing adult teeth may make you wonder why mammals are typically limited to two sets of teeth. Sharks famously replace their teeth throughout their lifetime, as do snakes and crocodiles. Wouldn't it solve a lot of issues if mammals did the same thing? This is certainly something I have pondered during visits to the dentist.

There is one marsupial (and only one) known to have evolved such a solution: the ability to endlessly replace their teeth.

The nabarlek is a tiny rock-wallaby with an apparently unlimited number of molars, which continue to appear throughout their lives as their teeth are worn down by the silica-rich plants that they eat.

Rock-wallabies are a diverse group of Australian macropods. Nearly twenty species are currently recognised, but some look very alike and have been subject to ongoing reassessment and redescription by taxonomists. New species have been described in most decades since the 1820s (most recently in 2014). They range in size from the minute monjon, which weighs just a kilogram (about 2¼ pounds) and measures 30 centimetres (1 foot) long with a tail a little shorter, to the 11-kilogram (24-pound) brush-tailed and yellow-footed rock-wallabies. Yellow-footed rock-wallabies are perhaps the most beautiful macropods. They have vivid yellow arms and legs, plus alternating rings of yellow and brown running down their long tails, with white flashes on their hips, flanks and faces.

Rock-wallabies fill similar niches to mountain goats in other parts of the world, and their agility over precipitous cliff faces and rocky escarpments is incredible. Watching them move at extraordinary speeds over implausibly complex and steep routes takes your breath away, particularly when you consider that it is all done by two-footed hopping, rather than the careful placing of one foot after another. Other than tree-kangaroos – which are rock-wallabies' closest relatives – macropods cannot move their feet separately on land (although they can while swimming, which they are surprisingly adept at). Large placental mammals that move apparently effortlessly in rocky terrain, like snow leopards, ibex and American mountain goats, have the benefit of four feet to keep them stable. Rock-wallabies are particularly impressive using just two at high speed.

Unlike most other macropods, rock-wallabies' tails are tube-shaped rather than tapering from a fat base, highly flexible, and many species have a tuft at the tip. This is critical for their acrobatic manoeuvres, acting as a counterbalance, even enabling them to switch direction in mid-air. Their ankle joints and ligaments are arranged to avoid rupturing upon landing after bounding over chasms, but this does mean that they store less elastic energy than their relatives on flatter ground.[18] As well as this, their feet are not as long as those of other macropods (which is a benefit on rough ground), they have a spongy underside to act as a cushion when they hit the rocks and their soles are rough, for added grip. The tree-kangaroos of the tropics have evolved similar features to help them exploit the bounty available among the branches in the rainforests of north Queensland and New Guinea.

On one trip to southwest Western Australia in 2010, I was extraordinarily excited about doing some fieldwork in Two Peoples Bay Nature Reserve, because of the fame of one of its macropod residents. I was officially there to help survey rakali (native water rats), but that wouldn't stop me from looking for what may be the world's rarest mammal.

Two Peoples Bay became famous among zoologists when the highly specialised noisy scrub-bird was rediscovered there in the 1960s, after being considered extinct for decades. Then, in 1994, another species returned from the dead in the same place: Gilbert's potoroo. It, too, had been presumed extinct.

For a few days before the rakali team arrived, I was the only person staying in the reserve, other than the park ranger, and I had the research field quarters to myself. I enjoyed the irony of being one person, alone in Two Peoples Bay. As my lift drove away, a rare quokka hopped out of the bushes, and it was later joined by more. Quokkas, the cat-sized macropods that those Dutch explorers met on Rottnest Island, have declined massively in mainland Australia since Europeans introduced cats and foxes, and the ease with which they could be seen that evening gave me hope for seeing one of their notoriously elusive relatives – the potoroos.

Potoroos and their relatives, the bettongs (collectively known as potoroids), are more petite than most other macropods, with shorter tails (though some are able to carry nesting material grasped in their curled tails) and shorter feet. They also have a more varied diet, typically digging for underground fungi and plant roots, and they'll also eat insects and fruit.

Sadly, potoroos and bettongs have suffered more than most groups from the changes brought about by European settlement. Foxes are a key culprit, but given the rarity of their rediscovered residents, these introduced predators are heavily controlled at Two Peoples Bay, and as I explored each night desperately hoping to see a Gilbert's potoroo I was amazed at the mammals I found. Not only were quokkas hopping around, but there were also plenty of quenda (the local bandicoots) and western grey kangaroos; and up in the peppermint trees were endangered western ring-tailed possums making little chirruping noises.

Although I would go on long hikes in the day (managing to hear, but not see, the noisy scrub-bird) and wander far at night with my spotlight, all of those marsupials could actually be found on the grass outside my window. It was a mammalogist's dreamland. The diversity *inside* the field quarters was impressive, too, with two carpet pythons living in the rafters, and, one day, I was quite concerned to spot a black, scaly tail poking out from under the fridge. There were instructions to keep the cabin doors closed to stop antechinuses coming in. As much as I would have been utterly thrilled to share my quarters with these little predatory marsupials, it did make me wonder whether the tail under the fridge belonged to a venomous tiger snake, which could have been attracted indoors to eat the antechinuses. I had a decision to make – either I left the owner of the black tail to its own devices and lived in ignorance about what was under the fridge, or I removed the doubt and confirmed what it actually was and whether I should be worried. The problem was that in order to identify it, I needed to lie on my belly with a torch and stick my face up close to the crack under the fridge, thereby cornering whatever it was with a rather vulnerable part of my body. Rather than spend all the time wearing boots and gaiters, I decided that ignorance wasn't bliss and that I needed to know. After some very tentative investigations, I was relieved to discover that my roommate was a half-metre-long (20-inch) King's skink – a chunky lizard. He was welcome to stay.

Sadly, I never set eyes on a Gilbert's potoroo, but given that there were only around thirty on the planet at the time, it was always going to be a long shot. However, a potoroid I am much better acquainted with is the burrowing bettong, or boodie, as it is known in Western Australia. It's such a common theme for Australia's native mammals that I feel like I've written this a lot already about other species, but burrowing bettongs were once widespread over much of Australia, however following European colonisation – and the cats and foxes they brought with them – their population plummeted and their range retracted heavily. Burrowing bettongs were wiped from the face of mainland Australia, and by the mid-twentieth century were found only on four islands off the coast of Western Australia. In the 1980s the species was then lost from one of those four – ironically the one that shared its name, Boodie Island – although happily it has been success-fully reintroduced there. Indeed, once introduced to fenced reserves on the mainland and to islands where cats and foxes are excluded, boodies can do very well.

Unless we are undertaking surveys for burrowing bettongs themselves, doing fieldwork where this species is present can be frustrating, as they can be so numerous and fearless that they obstruct work on other species. Faure Island, in Shark Bay, is within sight of the Western Australian coast and has been managed by the Australian Wildlife Conservancy since 1999. Before I went there, I had assumed Shark Bay got its name because some eighteenth-century explorer had spotted a shark there once and wrote it down in a journal. And indeed, when I read English buccaneer William Dampier's *A Voyage to New Holland*, an account of his trip from 1699–1701, he had written of the bay, 'Of Sharks we caught a great many, which our Men eat savourily. Among them we caught one which was 11 Foot long. The space between its two Eyes was 20 Inches, and 18 Inches from one Corner of his Mouth to the other'.*[19] It was Dampier who named the place Shark's Bay. However, the name isn't just some historical hangover: sharks continue to abound there in great number, and occasionally in great size, as in Dampier's day. While I was there, if ever we wanted to escape the heat and the flies (there were an extraordinary number of flies) and go for a swim, one of us would take watch to keep an eye out. Thankfully the water is crystal clear, so 11-foot sharks would be noticeable from a distance. At one point, while I was wading in the shallows to photograph cow-nosed rays, a small reef shark swam up to one and bit it in half, surrounding my legs in a pool of the ray's blood.

We were not there for the sharks, however. One of the key purposes of the trip was to ascertain whether the threatened djoongari could be captured in sufficient numbers to take to establish a new population in a mainland sanctuary. Djoongari are native rodents also known as Shark Bay mice, which – once again – had been driven to extinction on the mainland, and only survive on offshore islands with no introduced predators.

However, we were faced with two challenges that ultimately stopped us from catching a single djoongari. They were there, but we couldn't catch them. The first was the boodies. Faure Island is oblong, approximately 12 kilometres (7½ miles) long

---

* This quote continues quite brilliantly, with a story about what they found inside the shark: 'Its Maw was like a Leather Sack, very thick, and so tough that a sharp Knife could scarce cut it: In which we found the Head and Boans of a Hippopotamus'. Of course, there are no hippos in Australia, but Dampier goes on to describe a dugong's face and skull. It's surprising that he didn't recognise it as such, as he had found dugongs (which he had called manatees) on his previous voyage to the Western Australian coast aboard the *Cygnet* in 1688 – perhaps the one in the shark's belly was too decomposed.

and 6 kilometres (3¾ miles) wide, and we were four people surrounded by 20,000 boodies. Once the AWC had removed the cats and foxes and introduced some boodies, they boomed. Boodies are very curious, and every trap we put out – even the ones that were far too small for them – would be disturbed by the boodies trying to reach the bait balls of peanut butter and oats inside. We tried putting the traps inside hard plastic tubes to keep the boodies out, but this was only partially successful. Eventually, we worked out an arrangement of traps that were spaced out in a way that distracted the boodies and left the mice a clear run at reaching our bait, if they could be tempted to.

But the second challenge was that a couple of weeks before we arrived, a cyclone had dumped several years' worth of rain on Faure, which is otherwise a typical 'desert island'. With this injection of water, the plants had gone into overdrive. It was beautiful to see, but meant that there was so much food around for the mice that they didn't need to eat our bait, and so even when the boodies didn't tamper with our traps the djoongari didn't go in them.

Despite their habits as saboteurs of ecological fieldwork, boodies are thoroughly charming animals to be around. They are aggressive and communicative towards each other and make some amusing noises. One in particular is a squelching fart noise (made by their mouths), which can be imitated by noisily forcing a pocket of air from out under your top lip. I have often been startled by a boodie hidden in a bush that squelched as I walked past. They are also energetic quarrellers. While they are foraging close to one other quite calmly, suddenly the peace can be breached when brief fights break out and they engage in what is known as the 'boodie kick' (although other potoroids do similar things). The boodies will rapidly drop onto their flanks and kick out at an adversary with their powerful back legs.

Burrowing bettongs are the only macropods to regularly shelter in underground warrens, which on the mainland (when they originally lived there) could have over 100 entrances.[20] These empty burrow systems can still be found on the Australian mainland, along with the giant nests of stick-nest rats, as palpable reminders of its lost animal communities – one species, the lesser stick-nest rat, is extinct, while the greater stick-nest rat now survives only on predator-free islands and in fenced reserves. Bettong burrows can make walking a challenge, particularly in sandier soil, where they more readily collapse under a person's weight. Far more importantly, however, the digging habits of successfully reintroduced populations are vital for the proper functioning of the ecosystem as a whole, by helping to return nutrients to the soil.

*

Another pocket of Australia where cats and foxes have been removed and burrowing bettongs reintroduced is the pioneering reserve in central South Australia called Arid Recovery – and that is where I finally met another of the species to have topped my must-see list: greater bilbies.

A reasonable assessment of a bilby's appearance is that all of its appendages are somehow out of proportion. They move on all fours, and attached to their body – which is around the size of a rabbit – are extremely prominent donkey-like pink ears;* a very long, thick tail, which is grey at the base, black in the middle section and has a bright white tuft for the last half; a long, sharply pointed conical snout, tipped with a small pink nose; macropod-like long, narrow feet on hindlegs that lift their rump high off the ground; and thick, stout front legs with large claws on the middle three fingers for digging their

---

* Bilbies' size, shape and ears are sufficiently rabbit-like for a movement in Australia to have established the bilby as the local symbol of Easter instead of the bunny. Cadbury's (and others) sell chocolate Easter Bilbies and conservation organisations take the opportunity to highlight the importance of Australia's threatened native wildlife, and the damage caused by introduced rabbits to the country's ecosystems.

burrows and finding food. Their coat is ash-grey on the back and white on the belly, and unquestionably the softest, silkiest fur I have ever handled.

Despite this curious combination of characteristics, bilbies are in no way unattractive animals. In his epic 1935 account of an expedition to central Australia to locate the desert rat-kangaroo (now extinct) – one of the greatest literary travel accounts I have read, peppered with casual encounters with species that no longer exist – mammalogist Hedley Herbert Finlayson initially describes eating the 'tender if grotesquely shaped talgoo' (using the Luritja word for the greater bilby), but goes on to say:

> Remarkable amongst the smaller forms [of mammal] is the talgoo, one of the so-called rabbit bandicoots, which has carried a number of structural peculiarities to grotesque lengths yet manages to reconcile them all in a surprisingly harmonious and even beautiful whole. . . . The coat is one of the most beautiful amongst the marsupials: fine, silky, slate-blue, and quite like a chinchilla.[21]

Once again, since Europeans arrived, greater bilbies' range has contracted from covering around 70 per cent of mainland Australia to about 20 per cent, but rather than withdrawing into a single corner of remnant population, as many species have, greater bilbies survive in scattered populations in at least three states (Western Australia, Queensland and the Northern Territory, and they have recently been reintroduced to New South Wales). Sadly, however, a second species – the lesser bilby – disappeared entirely in the 1960s.

I had attempted and failed to find greater bilbies on previous occasions, both in the Channel Country near the border of Queensland and New South Wales, and sitting through the night watching burrow entrances near Broome in northwestern Australia. We eventually met on my second visit to Arid Recovery, and they didn't disappoint.

I had been watching a large burrow that had tracks and fresh diggings in the sand around it, and eventually one popped out, sniffed the air and then trotted off. At slow speeds, bilbies, like their relatives the bandicoots, bounce from the front feet to their back toes, much like a rabbit but with longer legs, flicking their rumps in the air as they go. At faster speeds it becomes more of an ungainly gallop. When some turned up in our survey traps on the subsequent mornings, I was able to see for myself how similar certain aspects of their hind feet are to those of macropods. Bilbies and bandicoots have the united (syndactylous) second and third toes, an elongated weight-bearing fourth toe and smaller fifth toe. However, since their teeth more closely resemble the dasyurids, it's unclear where bilbies and bandicoots fit in the marsupial family tree – are they closer to dasyurids or the group that contains macropods, possums, wombats and koalas? Some evidence also points to them being the closest relatives of the third major grouping of Australian marsupials – the marsupial moles. As such, they have caused taxonomists quite a lot of trouble over the years.

Most bandicoots are like less elaborate versions of a bilby: short legs, short tail, shorter ears and a shorter (but still long and markedly pointed) nose. They don't live in burrows but do dig little conical pits in order to find their food – invertebrates, tubers and fungi, as well as seeds and small mammals. The daintiest species – the threatened western barred bandicoot, which has also been reintroduced to both Arid Recovery and Faure Island – weighs just a couple of hundred grams. It is a grizzled grey with faint, light-coloured stripes running across its rump, with big eyes, big ears and a conical snout. Reaching more than ten times heavier, the northern brown bandicoots of Australia's north and east coastal regions are the largest members of the group.

*

I mentioned at the start of the chapter that, aside from platypuses and echidnas, marsupials only make up approximately half of the native land mammals in Australia, and before we move on from this celebration of the country's incredible fauna, I don't want to forget the placental mammals that make up the other half. Dingoes arrived in Australia around 4,000 years ago, transported on boats as semi-domesticated descendants of Asian wolves by sea-faring people. They subsequently spread across the whole of mainland Australia (but never reached Tasmania). As such, they are not considered native, but play a major role in Australian ecosystems and Aboriginal cultural life.

The bats and rodents in Australia have just as fascinating roles in the ecosystem as marsupials, and equally impressive adaptations. It's curious, however, that in the introduction to his epic three-volume *Mammals of Australia*, the legendary naturalist John Gould wrote:

> As regards this great country, it may be said that its most highly organized animals, if we except the Seals, are the various species of Rodents, and the equally numerous insectivorous and frugivorous Bats, both of which rank among the lowest of the Placentals.[22]

Through his books, Gould arguably did more than any other person to bring Australian wildlife into the public imagination of the West in the nineteenth century. The exquisite illustrations in these works continue to be the most commonly used depictions of some Australian species today. It is notable, then, that even Australia's native placental mammals are dismissed as being the 'lowest'.

Despite what Gould said, bats are unquestionably incredible animals. True flapping flight (as opposed to gliding) has only evolved three times in the long history of animals with bones: in the birds, in the pterosaurs (reptiles that went extinct at the end of the Cretaceous Period) and in the bats. Each of these groups have evolved different underlying structures for their wings. In birds, the wing involves feathers coming off of three fingers; pterosaur wings are a membrane attached to their fourth finger; and bats spread a wing membrane between four digits and across to the body and legs, and it often runs in between the legs and tail as well. The feet and legs of bats have evolved to allow them to hang from them at rest without requiring any energy to grip. Many species are famous for the ability to precisely locate objects and navigate complex environments by analysing echoes from sounds they produce. All in all, bats are extremely successful, making up over one-fifth of all known mammal species,[23] and in Australia they represent nearly a quarter of the land mammals – around eighty species.

Perhaps the most unusual Australian bat is the ghost bat. They are large (for a small bat), weighing 150 grams (5 ounces), with pale white fur, pinkish wings and a sizeable nose leaf. While most bats feed on insects, ghost bats eat small mammals, frogs and lizards, which they catch on the ground, and snatch birds and other bats out of the air. They have large eyes and big pink ears, which are fused above their head, and hunt through sight and sound as well as echolocation. Enveloping their prey in their wings, like a gothic depiction of a vampire wrapping his cape around his victim, they kill with long, sharp teeth and strong jaws.

Most bat species hunt insects rather than vertebrates, and one of the most effective ways of studying them is to position what's called a harp trap in likely-looking flyways where bats may pass through as they hunt, like at the bases of cliffs, over dry river beds or clear pathways through forest. A harp trap comprises two horizontal bars with an array of nylon threads strung between them. If a bat flies into the threads, it gently slides down them and falls into a soft canvas bag at the bottom, which has a flap that makes them unlikely to crawl out. Because bats have very high energy and water demands, these traps must be checked extremely regularly so that no bat spends much time in there.

One of the most surprising-looking animals I have ever seen was found in a harp trap. Orange leaf-nosed bats from northern Australia are one of the ghost bat's favourite foods. Their body and head are covered in thick, luminous pumpkin-orange fur, and their wing membranes are grey-brown with bright pink over their limbs. The nose leaves (which are thought to help direct sound for echolocation) occupy most of their face – from the top of their lips to well above their eyes – and resemble baby pink peeled walnuts. They can be found in roosts of around 20,000 individuals, which must truly be a sight to see.[24]

Australia is also home to several species of fruit bats, known as flying-foxes because their faces resemble those of foxes. They specialise in fruit, nectar and pollen, and as such are important pollinators for a lot of tree species. When a tree is in fruit, massive groups can descend and cause quite a spectacle. They are very noisy animals, and their size can be breathtaking. The largest Australian species has a wingspan of 160 centimetres (5¼ feet), which is almost exactly my own arm span.

The other large slice of Australian land mammal diversity is made up by rodents – comprising nearly a quarter of all species there (more than seventy species at the time of European colonisation, at least fourteen of which are now extinct). While by and large they retain the typical rat- or mouse-shaped body plan, they have diversified to occupy a wide range of roles in the ecosystem, and some of them are extremely beautiful.

The rakali is among the largest and is found across much of the country where permanent water can be found. In the east, they share their range with the platypus – Australia's only other mammal specialised for open water – but rakali are also found around salt and brackish water. Their lifestyle and adaptations are like otters elsewhere in the world. They have dense coats (which, like otters, were exploited by the fur trade), long white-tipped tails, golden bellies and partially webbed hind feet. Their eyes, ears and noses are positioned high on their heads so that they remain above the surface as they swim. In the water, rakali hunt pretty much anything they can catch, including fish, turtles, crustaceans, molluscs, frogs, aquatic birds and lizards, and they occasionally eat plants. They also hunt on land, where they might take small mammals or carrion.

Up in the trees, several different groups of large Australian rodents have evolved to fill the niches formed by squirrels and arguably some monkeys elsewhere in the world. The rabbit-rats, tree-rats and white-tailed rats are all adept climbers, either using trees for shelter or food or both.

Rodents owe a lot of their global success to the adaptation of their teeth. Their incisors – like a wombat's – are open-rooted, ever-growing and arranged in a way that ensures they stay sharp. Every time a rodent bites down on its food, or even just closes its jaws together, the top teeth grind down against the bottom ones, so that they sharpen each other. The layer of enamel on the front of their incisors is harder than the back, and in many species this front layer is strengthened even further by the incorporation of iron minerals. If you've ever noticed that some rodent incisors are orange, that's not down to poor dental hygiene – it's because the iron in their teeth is rust-coloured. Because of the difference in hardness of the two layers, when their bottom teeth bite against their top teeth, the softer back of the tooth grinds down faster than the harder front, which creates an ever-sharp cutting blade at its point. Because these teeth never stop growing, as the tip is worn down, it grows out from the base to maintain the same length. However, if they lose a tooth, its opposite number will keep growing in a loop out of their heads and can even grow round to spear them on the forehead or chin.

The giant white-tailed rats of tropical Queensland put these teeth to use cracking into tough tree nuts that no other species can open. Coconut shells with their distinctive gnaw marks

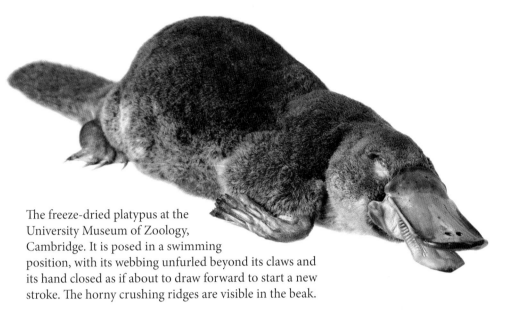

The freeze-dried platypus at the
University Museum of Zoology,
Cambridge. It is posed in a swimming
position, with its webbing unfurled beyond its claws and
its hand closed as if about to draw forward to start a new
stroke. The horny crushing ridges are visible in the beak.

How the platypus was introduced to Europe. This illustration by Frederick Polydore
Nodder appeared alongside George Shaw's first scientific account of the species in 1799,
in volume 10 of *The Naturalist's Miscellany*.

Corrie, the first platypus to be
bred in captivity, photographed
at about nine-weeks old in
January 1944 at Healesville
Sanctuary in Victoria, under
the care of David Fleay.

My first echidna encounter, in Tasmania, with the echidna hunkered down with its head and feet tucked away in its 'DEFCON 2' pose.

A Tasmanian echidna foraging, pushing its snout into the soil to search for invertebrate prey.

Stages of echidna egg development and a young puggle, before it has started to grow spines, preserved at the University Museum of Zoology, Cambridge.

Tasmanian devils are now the largest surviving marsupial carnivores.

A common wombat on a buttongrass tussock in the Tasmanian highlands.

The Kongouro from New Holland, painted by George Stubbs in 1772 for Joseph Banks, following encounters between kangaroos and the crew of the *Endeavour* when they were forced to beach the ship after it ran aground on the Great Barrier Reef in 1770. This was the first European painting of an Australian animal, and is now at the National Maritime Museum in London.

A western pygmy-possum rousing from torpor, on Kangaroo Island.

A long-tailed planigale demonstrating both the ferocity of the world's smallest marsupial, and the flatness of their heads. These tiny predators live in cracks in the soil – the name planigale means 'flat weasel'.

A dark-coated eastern quoll being released during fieldwork in Tasmania.

A young short-eared rock-wallaby in the Kimberley region of northern Western Australia. Rock-wallabies have shorter feet than other macropods and a tubular tuft on their tails, both of which help with their agility as they race across cliff faces and boulder piles.

A greater bilby emerging from its burrow at Arid Recovery wildlife reserve in South Australia.

Three baby Tasmanian devils attached to teats in the pouch at two to three weeks old. Female devils give birth to around 20 babies, but as they only have four teats, the maximum number they can raise at a time is four, because marsupial infants initially attach uninterrupted to a teat, so it cannot be shared.

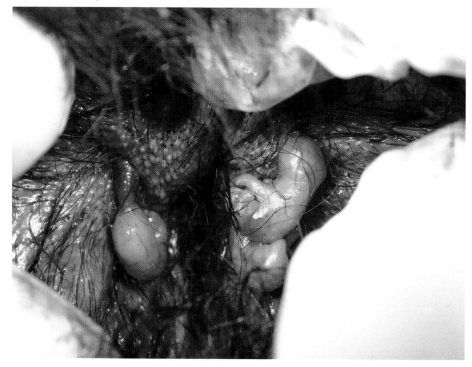

Three baby devils suckling in the pouch at around four months old. At this stage, the mother will start to leave her young in a den while she goes out to forage.

A striped possum in North Queensland. Its elongated fourth finger – which is used to hook beetle larvae out of holes in wood – can be seen wrapped around the branch. This adaptation has evolved independently in both striped possums and aye-ayes.

A female thylacine and one of her juvenile young, photographed in 1903 at the National Zoo in Washington, D.C. When they arrived at the zoo in September 1902 (along with two of the juvenile's siblings), they were the first of nine live thylacines to be seen in America.

The largest marsupial to have ever lived was the rhino-sized wombat-relative *Diprotodon optatum*, which died out around 50,000 years ago. This skeleton cast is on display at the University Museum of Zoology, Cambridge. A southern hairy-nosed wombat skeleton stands at its feet, to give a sense of scale.

A taxidermy echidna at the Grant Museum of Zoology, UCL. As is very common for echidnas in museums, this specimen is not echidna-shaped. Its hind feet should point backwards, not forwards, and its belly should not be flat to the floor like a frog.

Western barred bandicoots were recently discovered to have comprised five separate species, four of which are now extinct thanks largely to feral cats and foxes. One species survived on islands in Shark Bay, and has since been introduced to a handful of predator-free reserves, including Arid Recovery in South Australia.

on them can often be found strewn across the beaches. A resident of the Daintree Rainforest once told me that when their house flooded it washed the labels off the cans of food in their pantry, disguising what was inside. One night, a white-tailed rat got in and chewed its way into one can to feast on its contents – it was coconut milk. This demonstrates the extraordinary strength of their teeth and jaws, but also suggests that they have an incredible sense of smell, being able to detect the coconut through the solid metal, as they left all the other cans, which contained foods far less favoured by this species.

Strength is also an attribute of some of Australia's more miniature rodents. There are a wide number of native mice across all Australian habitats, including the super-agile hopping-mice, which have evolved similar characteristics to the jerboas of Africa and Asia and the kangaroo rats of North America (which, incidentally, are one of the few non-Australian animal groups which are named after an Australian species). Despite only being 10 centimetres (4 inches) long, they can hop over 60 centimetres (2 feet) with their powerful back legs and dig burrows a few metres long. H. H. Finlayson wrote of them:

> They average about twice the bulk of a house mouse but are built on totally different lines, with elongated hind limbs, very short forelimbs, long ears and a long, bushy tail. When undisturbed and feeding quietly, they go on all fours, but when startled, they take to their hind toes and bound like tiny kangaroos. The legs are so thin that when moving rapidly they are almost invisible, and the thing has rather the appearance of a ball of down being blown along, rather than an animal under its own motive power.[25]

Australia is home to a few animal architects, too. The minuscule pebble-mice weigh 10 grams (⅓ ounce) and yet construct rock piles around the entrances to their burrows that can cover 10 square metres (over 100 square feet). The exact purpose of these mounds is unclear, but it is something they put a lot of effort into arranging. The mice can move individual pebbles that are equal to their own body weight, and the combined mass of all the pebbles in a mound may be thousands of times heavier than the inhabitants of the burrow it covers.[26]

The construction award, however, surely goes to the greater stick-nest rat. These beautiful, round-faced rodents with short noses and large ears gnaw dead sticks to size and weave them together to form massive nests. Reaching a metre (3 feet) high, and 2 metres (6½ feet) wide, these are often built around other structures such as bushes, but in the Arid Recovery reserve where they have been reintroduced following another mainland extinction, they can be seen around fences, within piles of tires and even under the toilet building. The rats glue their nesting material together with their urine, which ensures they are solidly built to provide protection against the elements as well as predators.

So, while most of Australia's placental mammals appear relatively familiar on the surface – in that, with a few exceptions, they are largely the same shape as the mice, rats and bats found across the rest of the world – many are nonetheless remarkable in their behaviours. Similarly, the general notion of how marsupials make more marsupials might seem familiar to us – everybody knows kangaroos have pouches – but as we will see, the specifics may be surprising.

8

Roo-production

The marsupial reproductive strategy requires an almost unbelievable feat from their minute newborns. The minuscule babies must climb, arm over arm, the significant distance from their mother's vaginas (yes, that's right, plural) to the pouch, unaided, and then fix themselves to an empty teat. This may explain why there are no marsupial bats or whales: the need to climb into the pouch means that they cannot evolve away from having hands with grasping fingers (and hence form wings or flippers*),[1] and the babies of fully aquatic marsupials would drown in their pouches. (The parachute-like membranes of the various gliders are wonderful evolutionary solutions for becoming airborne without requiring significant modifications to the all-important grasping hands.)

One departure to this model is seen among the bandicoots, whose young don't crawl to the pouch – the mother curls over and they sort of flop out into the pouch opening. Given that newborn bandicoots don't start life with a marathon climb, this birthing technique has potentially freed them of the evolutionary constraint that is applied to most other marsupial hands. And indeed, some bandicoots are among the few marsupials to have done something different with their hands. The

---

* However, the three smaller fingers on a pterosaur's hand could form a grip (the wing is formed by the fourth finger), so if marsupials were to evolve a wing like a pterosaur, perhaps they could climb up their mothers' fur.

pig-footed bandicoots have toes that resemble miniature trotters, with long, slender front legs tipped with little hooves. This is intriguing. The evolution of hooves sparked a massive diversification in placental mammals (think of how many species of antelope, deer, cattle, goats, sheep, giraffes, camels, horses, rhinos, tapirs and pigs there are). Could pig-footed bandicoots be the start of a similar explosion in marsupials? Sadly, we will never know, as the only two species disappeared by the 1950s, driven to extinction by introduced cats and foxes. The second of those species – *Chaeropus yirratji* – was only described from museum specimens in 2019, demonstrating the value of historic collections in understanding the scale of biodiversity loss over recent centuries.[2] We have one of the eighteen known specimens of this new species in the University Museum of Zoology in Cambridge, which I consider a profound responsibility. The eighteen specimens, held across just five museums worldwide, are the only known physical evidence that the species ever existed.

Pig-footed bandicoots are now extinct, but are the only marsupials to have evolved hoof-like digits, as can be seen in this illustration from Museums Victoria by Gerard Krefft, based on specimens from an 1857 expedition to north-west Victoria. Krefft went on to become director of the Australian Museum.

Aside from the lack of crawling in their infants, there's another difference in bandicoot reproduction. Although placental mammals get their name from forming a placenta that nourishes their growing embryos in the womb, marsupials do also form one. In most species this is simply a means by which the embryo's yolk sac (which 'feeds' the foetus) gains nutrients from the mother across the wall of the womb. However, in some species – including bandicoots, koalas and wombats – a complex integrated placenta forms to exchange materials between the mother and young in a similar way to so-called placental mammals.[3] Despite this, bandicoots actually have one of the shortest pregnancies of any mammal species – at just twelve and a half days for the long-nosed and northern brown bandicoots[4] (some species of dunnart may just beat them to the record, at eleven days). The newborn bandicoots' umbilical cords remain attached to them when they first reach the mother's nipples, running from the vaginas to the pouch as taut strings.

I should explain this plural vaginas situation. Like humans (and all other mammals), female marsupials have two ovaries and two oviducts. In placental mammals, these lead to a single uterus, which has a single cervix and vagina leading to the outside world. In marsupials, by contrast, there is a right uterus and a left uterus, each with its own cervix and vagina. When marsupials become pregnant for the first time, they then grow a third vagina leading from the point where the two uteri meet down to the cloaca (most marsupials have a single opening, like monotremes, for all their birthing, mating and waste-disposal needs). In some species, this third vagina then fuses up again after birth, until their next pregnancy, while in others – including kangaroos and their relatives – it remains in place throughout their lives.

Despite only having been in the womb for a matter of days or weeks – and therefore resembling minute pink jellybeans – newborn marsupials already have extremely well-developed arms and lips. That's what enables them to climb and suckle respectively. It is astonishing that such tiny babies have the requisite musculature, skeletal strength and nerve-control to allow them to achieve this athletic feat. They can't see or hear, and while we don't know exactly how they navigate the route, they can smell at this young age, so perhaps that's what leads them to finding the teat.

Kangaroos are born after four to five weeks' gestation, weighing just 0.8 grams. They eventually grow to become the largest living marsupials at over 90 kilograms (198 pounds). At the other end of the scale, honey possum newborns weigh less than 5 milligrams: one quarter the mass of a grain of rice. This makes them the smallest known of any mammal. Honey possums hold some other reproductive records too. They are believed to have the largest testicles relative to their body size (making up nearly 5 per cent of their body weight) and the longest sperm, in absolute terms, of any mammal in the world. It is unusual for species as small as a honey possum to have differently sized males and females, but in their case the females are almost twice as big. They carry their young in their pouch until they reach a more advanced stage of development than most small marsupials (the total weight of a litter can almost match the mother's own weight), which may explain why the females are so much bigger than the males.[5]

Today, perhaps the most common way of describing how marsupials reproduce – in journal articles, museums labels, popular science books and TV shows – is to say that they give birth to 'underdeveloped young'. What people mean is that marsupials have very short pregnancies, because most of their infant growth takes place by suckling milk from a teat rather than in the womb. As such, they are born at an earlier stage of development than placental mammals. However, the word *underdeveloped* implies that they come out too soon – that they have somehow failed to reach the level of development that placental mammals have. It's another incorrect subconscious suggestion of inferiority. Marsupial babies are born precisely as developed as they should be. They are *less* developed at birth compared to placentals, but they are not *under*developed. There is no hierarchy – they are different but equal.

Marsupials absolutely fascinate me, which is why I spend much of my life chasing them around Australia. It is hard to describe the amazement when working with Tasmanian devils of looking into their elastic pouch and finding four perfect mini-devils, each the size and shape of a mango (considering that a large female devil is about 8 kilograms/18 pounds, by the time her four youngsters have reached mango-size their back-ends are sticking out of the pouch). Or of catching a northern quoll whose eight babies attached to her teats are so long that they stick out of the pouch (actually more like a small fold of skin in this species), causing her to run on tiptoes to avoid dragging them along the floor.

When we look at reproductive tactics across the animal kingdom, there are different strategies for ensuring that at least some offspring reach maturity. Many animals – like a great deal of fishes, turtles, the frogs in our garden ponds and countless invertebrates – produce large numbers of young and largely leave them to fend for themselves. They are playing an odds game: if they have very many babies there's a good chance that a few will make it through the trials of infant life, even without parental care. In other species, including mammals and birds, the tactic is typically the opposite: produce small numbers of babies, but invest heavily in raising them. Overall, there is a trade-off between numbers of young and how much parental care they are given.

Some marsupials, including devils and quolls, make the most of both of these tactics – one after the other – which I think is a stroke of evolutionary genius. Because marsupial babies are so tiny, by the time they are born they have cost the mother very little in terms of energetic investment. An eastern quoll, for example, is born after around three weeks in the womb and weighs just 12.5 milligrams. Because these babies 'cost' so little, the mother can give birth to up to thirty of them at once, with a total weight of significantly less than half a gram. Devils do something similar, producing up to twenty newborns that are each just 6 millimetres (¼ inch) long, with a total litter weight of less than 4 grams (¹⁄₁₀ ounce). For an adult devil, that's nothing.

However, there is no chance that they can raise anywhere near that number of babies, as eastern quolls only have six teats, and devils have four. When baby marsupials first attach to a teat, the tip swells inside their mouths and attaches the young to its mother for the first chunk of their infancy, uninterrupted. This means it is impossible for marsupials to raise more babies than they have nipples. At least for the first weeks in the pouch, the teat actively pumps the milk into the baby, so it doesn't need to suck.

The devil and quoll strategy is to initially play the odds by having a significant overabundance of babies. This maximises the chance that four or six, respectively, will make the gruelling journey to the pouch. And once the 'winners' are in there and start growing by suckling milk, the mother begins investing the high levels of care that makes it likely that they will survive to weaning (none of the others, who didn't get hold of a teat, can survive). Even with that number starting the race, we very regularly find nursing mothers that do not have all their teats occupied, which goes to show just how difficult the journey must be.

This demonstrates that the marsupials' pouch-based system is certainly not primitive nor inferior to what we placentals do, which is often what people have implied (John Gould described kangaroo reproduction as 'a very low form of animal life, indeed the lowest among the Mammalia, and exhibits the first stage beyond the development of the bird').[6] The marsupial method is a fantastically sensible strategy for raising young in an unpredictable environment. If food or water become insufficient, the mothers are stressed or in fear of their lives, or the babies otherwise become a dangerous hindrance, it is very easy (if perhaps emotionally callous to our human sensibilities) to put a halt to proceedings. Marsupials can simply eject the young before too much energy has been invested in them. This is far harder and more biologically costly to do from within a womb. Plenty of accounts of colonial kangaroo hunts mention kangaroos removing their joeys from the pouch during the chase, presumably to increase the mother's chance of escaping.[7] There are some species that empty their pouches so readily that ecologists studying them have to take extra care to ensure babies aren't abandoned. When working with bettongs, for example, we have to use low-tack tape to temporarily close up their pouches, so the babies aren't left behind when we release the females.

If conditions are good, the system the kangaroos and wallabies employ allows for a constant conveyor belt system of baby production. At any time, they can have three generations in care at once. The eldest would be a large 'young-at-foot', which is still suckling but is too big to fit into the pouch – it may do a bit of grazing or browsing, but will periodically stick its head into the pouch and suckle from the same teat it has used since it was born (only now the teat is far longer and wider than when the infant first attached – the teat grows with the joey's mouth). The next infant would be a tiny newborn, physically attached to a different, far smaller teat which produces a different kind of age-appropriate milk (that's right – kangaroos can produce different milks from different teats at the same time). And the third generation would be an early-stage embryo, lying dormant in the womb, awaiting a teat to be freed up. Kangaroos typically mate again in the hours after giving birth, but the resulting embryo is then held in stasis. Once the young-at-foot is weaned, the mother's hormones change and kick-start the development of the embryo. Some macropod species combine this delayed embryonic development (called diapause) with seasonal triggers for restarting the baby's growth, like day length, to ensure they have enough food while raising young.

In 2020, scientists discovered that swamp wallabies take this strategy even further, by ovulating, mating and fertilising a second embryo in one of their uteri when they already have an earlier developing embryo in the other. This means that female swamp wallabies can become pregnant twice at the same time, before giving birth to the older embryo. Throughout their reproductive lives they are never not simultaneously pregnant and lactating.[8]

*

Other species of marsupial do the exact opposite to breeding many times continuously over the course of their lives – they squeeze all their breeding into a single, explosive event, and then die as a result of their exertions. This strategy is known as suicidal reproduction (or more scientifically, semelparity) and is the kind of biological story that the press enjoys getting their teeth into. Headlines for stories on this topic have included, 'Doing it to death', 'Why a little mammal has so much sex that it disintegrates', 'Sex, sex and more sex, and then death', 'These newly described marsupials basically sex each other to death', and 'What a way to go! Male marsupial found to sex itself to death after intensive 14-hour mating sessions in its final fortnight'. All of them are accurate.

Among mammals, suicidal reproduction is only found in marsupials: in some of the Australian dasyurids (carnivorous marsupials) and a couple of South American opossums. No male antechinus, phascogale or kaluta will live to see their first birthday, and the same is true for some populations of northern quolls and dibblers. What happens is that all the females become receptive simultaneously, during a very small window in time. At that point, the males more or less forego eating and hardly sleep in order to dedicate their efforts to fighting other males, hunting for females and then copulating frantically with as many of them as possible for up to fourteen hours at a time. These efforts can see males shrink dramatically – losing up to half of their body weight.

These events are mainly caused by high levels of testosterone and cortisol in the males' blood. Cortisol is a stress hormone, which causes proteins in the animal's body to break down into sugars, which fuels their frenzied sexcapades, but also significantly compromises their immune system. As they consume their own bodies, parasites and diseases take hold. It doesn't take long for their internal systems to collapse, causing their death in rather unpleasant ways.

In all animals, investing resources in producing and raising young comes at a cost to investing in one's own body and long-term survival (as I'm sure many human parents would agree). Generally, these trade-offs play out over an extended period throughout the animal's entire adult life. The suicidal reproducers have taken this compromise to the extreme, sacrificing literally everything for one breeding event. 'Live fast, die young' is certainly the way of things for these species. However, that maxim typically ends, '. . . and leave a good-looking corpse'. This is definitely not what happens here. I've caught male northern quolls towards the end of their short breeding window, and they are balding, covered in scabs, sores, ticks and other parasites, and are pretty darn disgusting.

The males have evolved in this way in response to an evolutionary benefit to the females. It is advantageous to the females to coincide the time at which they need the most energy – when they have growing young in the pouch – to the point in the year when the most insect prey is around. And so they breed in advance of this and come into heat only for a brief period in the year, all at the same time. This has a knock-on effect of increasing competition among males – if they don't mate in that short window, they don't mate at all. It not only requires them to put energy into fighting other males and wooing females, but also brings about sperm competition. This is an evolutionary pressure that can involve increasing the odds of fertilising a female's eggs by producing more sperm each time they have sex than they would need to if the females weren't promiscuous. Females drive this further by mating with many males during this short window.

Sperm competition drives the evolution of larger testicles to produce more sperm, and correspondingly high levels of testosterone, which reduces the males' survival by compromising their immune system. Essentially, the whole situation is driven by the females – they have brought about the modification of the males' biology in order to improve their own reproductive success.

As mentioned, all this mating takes place around a month before the point in the year when their invertebrate prey is most plentiful. So when they give birth and are lactating, it is at its peak. Not only that, but by then the males are all dead, so not only is the bounty in full swing, but the females can enjoy it all to themselves without the males to compete with them for food.[9]

*

When females mate with many males they are gaining evolutionary benefits as it can increase the odds that at least some of their young will have the genetics to grow up successfully and reproduce themselves. Thinking about it the other way, if females only mate with one male, they are basically – and literally – putting all their eggs in one basket. If that male is a dud, then their young will not succeed.

As a result, females of many species often mate with many suitors (and thereby bring about the sperm competition described above). In those that then give birth to more than one infant, different babies in the same litter can have different fathers. This can have remarkable outcomes for eastern quolls, which are unusual among wild mammals in that they come in two very distinctive colours – either a golden-fawn with white spots, or dark chocolate-black with white spots. We can catch mothers with young of both colours in their pouch, from different fathers. A study from 2020 on suicidally reproducing northern quolls found that all eight of the babies in one litter could each have different fathers.[10]

Feather-tailed gliders are another marsupial that have litters with young fathered by more than one male. This species can also be highly social, with communal nests of up to twenty-five individuals. For many marsupials, once the young are too large to be carried in the pouch, the mother leaves them behind in a den while she goes out to forage, and then comes back to allow them to suckle. In the feather-tailed glider, it seems that the females without young may stay with the babies while the mothers are out feeding. Not only that, but when they return, they often have babies that aren't their own suckling from their teats.[11]

Marsupials are most famous for their pouches, even though not all species have them, but in truth, as astonishing as the details of raising a minute infant on the *outside* of the mother's body are, there is so much more to marsupial reproduction than that.

9

The Missing Marsupials

All in all, the diversity of its mammals is one of the things that draws me to Australia's wildlife. You can travel just 50 or 100 kilometres (30 or 60 miles) and find that the rainfall, geology, soil chemistry, plants and habitat type have all changed, and so the community of animals you find there has changed, too. However, the diversity we see today – or even what the first European colonists found in 1788 – is only a paltry remnant of what it was in the recent geological past.

The Pleistocene Epoch is a period of the Earth's history ranging from 2.6 million to 11,700 years ago. It is characterised by a series of cold or glacial periods interspersed by warmer interglacial periods, which gives it its alternative name: the Ice Age. Worldwide, the Pleistocene is famous for its extinct giants, including 10-tonne (11-ton) mammoths, 4-tonne (4½-ton) giant sloths, giant deer with 4-metre-wide (13-foot) antlers and short-faced bears that approached 2 metres (6½ feet) at the shoulder when standing on all fours.

At the very centre of the University Museum of Zoology in Cambridge, where I work, is a cast of the rhino-sized skeleton of *Diprotodon optatum*: Australia's very own Ice Age giant. This 2.5-tonne (2¾-ton) relative of the wombat is the largest marsupial to have ever lived. It was thirty times heavier than a large male red kangaroo. When I look at this specimen – as well as the only real fossilised skeleton of a South American giant sloth on display in the UK, which stands alongside it – I am perpetually disappointed by the fact that we missed out on sharing the planet with these beasts by what is in geological terms just a blink of an eye. *Diprotodon* disappeared less than 50,000 years ago, and giant sloths only 10,000 years ago. Both of these dates shortly follow the first appearance of humans on their respective continents.

Phascolonus gigas ('gigantic pouched ass') was a true wombat – a close relative of today's species, even more so than *Diprotodon*. Unlike *Diprotodon*, *Phascolonus* appears to have burrowed, as modern wombats do. At 200 kilograms (440 pounds) and nearly 2 metres (12 feet) long, that would make it two-and-a-half times heavier than the largest living burrowing mammal today, the aardvark.* *Phascolonus* was around ten times larger than today's wombats (though similarly shaped), and *Diprotodon* was more than one hundred times larger. Some palaeontologists have described *Phascolonus* as potentially the largest burrowing animal to have ever lived;[1] however, there is evidence that the 4-tonne (4½-ton) giant sloths may have burrowed too, as tunnels in Brazil have been found with massive claw marks in the walls, apparently from the huge creatures that dug them. In any case, the existence of a wombat ten times larger than modern species constructing burrows is intriguing. In 1960, a 15-year-old schoolboy called Peter Nicholson became famous for being the first person to map and scientifically describe wombat burrows. At night, he would go 'wombatting' – sneaking away from his boarding school on his own (so no one knew where he was, should anything have gone wrong) and assiduously squeezing his way a significant distance down into their tunnels with pegs and a ball of string.[2] His findings were published in his school magazine, and the article remains one of the most useful studies on wombat burrows ever written. Following his school exploits,

---

* Unless you count the fact that polar bears burrow into snow, which is obviously cheating.

Nicholson did not go on to become an academic wombatologist – instead, young Peter studied economics.

Most adult humans would find themselves too large (and perhaps insufficiently brave*) to follow Nicholson's shimmying and journey into a wombat's world so intimately, but presumably it would have been an easy crawl to explore a *Phascolonus* burrow.

I'm often asked whether the massive *Diprotodon* could have produced square scats like its smaller relatives, but I suspect they didn't. Common wombats produce cubes, but hairy-nosed wombat poos are less geometric (as some studies position *Phascolonus* as most closely related to the hairy-nosed wombats, they probably didn't either). Koalas make lozenges, and they are the next closest relatives to wombats after the diprotodontoids and some other fossil groups. This suggests that poo cubes only evolved with common wombats.

Alongside a range of differently sized species of *Diprotodon* were a whole host of incredible Pleistocene Australian mammals that would have added to the wonderful diversity we see today. In that same group of giant wombat relatives there were beasts called *Zygomaturus*, which as well as having prominent bony projections on their cheekbones, may have had horns on their sizeable snouts, like a rhino. Some palaeontologists have suggested that they may have lived like hippos,[3] while others think this is rather fanciful.

A further closely related group included the half-tonne (1,100-pound) *Palorchestes*, which are often called 'marsupial tapirs', but they were more like cow-sized giant sloths than tapirs. They probably had long tongues like a giraffe and small trunks. And in New Guinea, their relatives included 150-kilogram (330-pound) marsupials called *Hulitherium*, which have been compared to giant pandas eating bamboo (again, that analogy is not universally accepted by all palaeontologists).

---

* DO NOT TRY THIS AT HOME.

The wombat relatives (collectively known as the vombato-morphians) were so diverse that they also included Australia's most effective marsupial predators: *Thylacoleo*,* the 'pouched lion', has been described by Australia's most distinguished palaeontologists as 'the most specialised mammalian carnivore to have developed *anywhere* in the world'.⁴ It was Richard Owen – the man at the centre of the platypus egg-laying debate – who first described *Thylacoleo* fossils (and *Diprotodon*, too). In 1859, he said that, based on its teeth, 'we may infer that it was one of the fellest and most destructive of predatory beasts'.⁵ More recently, it was found to have had the highest bite force relative to body weight of any mammalian predator to have ever lived, suggesting it could take down very large prey – possibly even a young *Diprotodon*.⁶

Thylacoleo weighed well over 100 kilograms (220 pounds), had an enormous, elongated bladelike tooth on each jaw and massive retractable claws and opposable thumbs, suggesting they were very capable climbers. This has led to only half-joking suggestions that Indigenous cultural memory of this species is in fact the source of the story of the 'drop bear', a supposed hoax tale of a vicious creature that jumps down from trees to attack passers-by walking below, told to terrify gullible tourists.

I spend a lot of my work and leisure time in museums and galleries and have seen some of the world's most celebrated

---

* A certain kind of person – myself included – enjoys the occasional absurdity of nested taxonomic group names. The term Vombatoidea is given collectively to the wombats, the Diprotodontidae (including *Diprotodon*, *Zygomaturus*, *Hulitherium* and their relatives) and the Palorchestidae (*Palorchestes* and its relatives). If you add these vombatoids to *Thylacoleo* and its relatives, the group is called Vombatomorphia. And if you add the koalas to that group, it becomes the Vombatiformes. Explained another way, Vombatiformes is the largest, most inclusive of these groups. Take away the koalas and you get the Vombatomorphia. And without the marsupial lions, you're left with the Vombatoidea. As my colleague Darren Naish put it, marsupial lions are non-vombatoid vombatomorphian vombatiforms (Naish, 2011). Confused? Sometimes I think taxonomists are doing it on purpose.

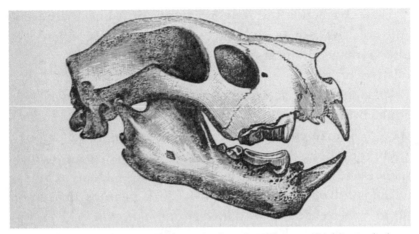

The skull of *Thylacoleo carnifex* – also known as the 'marsupial lion' – had a long, bladelike tooth on each jaw for slicing flesh, much like the carnassial teeth seen in big cats and other carnivorous placental mammals. It is an example of convergent evolution.[7]

artworks. However, I have never encountered anything in a museum that evoked as profound an emotional reaction as an artwork painted in ochre on a cliff in Kakadu National Park in the Northern Territory – it was one of a handful of known depictions of a thylacine dating from before their extinction in mainland Australia, which occurred 3,200 years ago at a similar time to the disappearance of mainland 'Tasmanian' devils.

Seeing that thylacine rock art was at least as exciting as many of the encounters I have had with live animals in Australia. But I can't imagine what it must have been like to come across a painting of *Thylacoleo*, which has not been seen alive for potentially 46,000 years. In 2008, a painting was discovered near the north Kimberley coast that almost certainly depicts this extinct predator. Unlike the thylacine, the animal in the artwork is given a rounded head and the depiction highlights the animal's massive forelimbs and prominent claws. It has stripes from its shoulders to its rump, but not on its thin tail. The painting is so detailed that you can even make out the genitals and what may be its tongue or its two protruding lower incisors.[8]

These were not the only predators prowling Pleistocene Australia. As well as thylacines and 'giant' devils,* there was also a range of meat-eating macropods, including a giant relative of today's diminutive musky rat-kangaroo, called *Propleopus oscillans*, which was around the size of an average adult British woman (70 kilograms/154 pounds). Today's species is small enough to easily fit into a child's lunch box.

On the whole, the Australian fauna included larger animals than we see today. There were metre-long (3-foot) echidnas ('*Zaglossus*' *hacketti*) and short-faced kangaroos that stood up to 3 metres (ten feet) tall (*Procoptodon goliah*). This was the largest known kangaroo ever, approaching a quarter of a tonne (over 500 pounds). Representatives of some of the same species we still find today were also larger than their modern forms.

If we were to look further back in Australia's deep history, beyond the Pleistocene, we would find a whole range of thylacine species and more cat-like predators related to *Thylacoleo*, including the leopard-sized *Wakaleo*, and *Priscileo*, one of several tree-dwelling predators. There was also a kangaroo relative known affectionately as 'fangaroo' (*Balbaroo fangaroo*) due to its massive canine teeth, but it was a herbivore. Its fangs are assumed to have been used in sexual selection, like mouse-deer and musk deer today.

*

Well over fifty mammal species had disappeared towards the end of the Pleistocene Epoch in Australia – if they were still around today, they would swell the country's land mammal count to around 400 species. But sadly, they are not. The reason why has been deeply controversial. A debate over whether humans or climate change wiped out the Australian

---

* Well, they were 15 per cent larger than living Tasmanian devils.

megafauna has been raging on for decades. Reasonable arguments have been used to implicate both factors. However, some archaeologists who doubted that human hunting caused the extinctions have suggested that Aboriginal people were not capable of having a significant impact on their land. This is an example of the kind of bigoted, unscientific assumptions that wrote off Indigenous Australians as 'primitive', a notion that was used to justify the dispossession of Aboriginal people by European colonists. Similarly, any naive assumptions that Indigenous Australians could not have altered species' survival because they were 'in harmony with nature' is as problematic as it is patronising.

As potential evidence for the climate change theory of extinction, one study on the chemical make-up of *Thylacoleo* teeth suggested that the animals lived in forest environments, which would have declined as the continent dried in the Pleistocene. What extinct herbivores ate can be inferred by testing their fossilised teeth, because browsing leaves creates a different carbon signature to grazing grass. This chemical marker is then passed on to the teeth of the predators that ate the herbivores – predators have the same carbon type in their teeth as their prey. The *Thylacoleo* teeth in the study carried the carbon signature of a browser, which suggests they hunted in forests.[9] This could imply that the decrease of forest habitats as a result of climate change led to *Thylacoleo*'s extinction. However, all the fossils in the study were from the same part of eastern Australia, which was indeed thought to have been forested at the time, but other fossils are also known from drier parts of Australia.

In his brilliant book *Australia's Mammal Extinctions*, which explores the last 50,000 years of losses, ecologist Chris Johnson points out that the species that disappeared in the Pleistocene were not the ones that should have gone extinct if the drying-out of the continent were to blame. As Australia cooled and dried, dense forests and other habitats with lots of diverse groundcover shrank. With those changes, we should expect small herbivores that rely on a diversity of high-nutrient plants and possums that glide between trees to be the most affected. Also, as the Ice Age peaked, those less tolerant of the cold – such as the tiny carnivorous marsupials – should also have suffered. But few of these animals are among the hordes that became extinct, Johnson argues. In fact, as many of the animals headed towards extinction, the favoured habitats of those that disappeared were becoming more widespread.[10]

The first people arrived in Australia at least 60,000 years ago. One potential problem with the idea that they quickly hunted these animals to extinction is that there is hardly any archaeological evidence of the extinct species being killed or eaten. But, then again, we wouldn't necessarily expect there to be. First, there is very little bone of any kind from the time of the extinctions, from animals that died of natural causes or otherwise. Second, there is little archaeological evidence that people hunted smaller animals either, when there is no doubt that they did. Third, as we shall see, very few animals would need to have been killed for them to become extinct, so there was very little chance of archaeological evidence being preserved for tens of thousands of years.

Johnson considered the expectations for what patterns we might see if people were responsible for the disappearance of the megafauna. The first prediction is that the species that reproduced slowly should have died out, as populations wouldn't have been able to replace the individuals lost to hunting. And they did. Also, the animals that were more accessible to hunters should have disappeared – those that lived down on the ground rather than up in the trees, or in more open habitats rather than in dense forest. And they did. Indeed, species that survived but have low reproductive rates live up in the trees.

Perhaps the most persuasive argument in favour of the overhunting theory is Johnson's mathematical model for just how few animals Aboriginal hunters would have had to have killed to wipe them out. The answer is not very many at all. His model involved designing a theoretical population of 1,000 *Diprotodon* with a set of conservative, sensible estimates for the birth rates and survival rates for differently aged animals in the absence of people. He then incrementally increased the death rate, to mimic hunting by people, to work out how many animals would need to be killed for the death rate to overtake the birth rate. Based on the evidence of how many people lived in suitable *Diprotodon* habitats at the time, his model showed that each person would only have to kill one animal every five years – or two animals a year per group of ten people – for the population to die out in just 520 years. The population only needed thirty animals to be killed per year for them to go extinct in the blink of a geological eye.

Johnson then refined the model based on the likelihood that people would probably avoid the largest animals due to the difficulty of killing a 2.5-tonne (2¾-ton) giant with a spear. If humans only targeted young animals in a population of 1,000 *Diprotodon*, thirty-eight killings a year would see them gone in 730 years. Based on these low numbers, the argument that the first Australians caused most of these extinctions is very appealing.

This hunting also explains the fact that many of the species that have survived up until today, from kangaroos to devils, are now smaller than they were in the Pleistocene. Larger animals typically start breeding later in life, and less frequently. As such, once humans started hunting in Australia, there would have been evolutionary pressure to decrease in size, with the result that the dwarfed versions that survived would breed earlier and more regularly. That's how we got to the animals we see today.

However – and it's a big however – we don't actually know with much confidence accurate dates for the extinction of these species – and they don't appear to have all disappeared at the same time. It's quite possible that some had died out before humans arrived in the region, and there is evidence that the last cold period was particularly severe, potentially pushing some species towards extinction. While there may be rock art depicting *Thylacoleo*, there is very little definitive evidence for people overlapping with most of the extinct species in Australia, but it is extremely unlikely that the last representatives of a declining species are preserved in the fossil record. This makes the precise timing of an extinction very difficult to pinpoint. Proponents of the idea that climate change killed off the giants are eager to point out that the most recently dated fossils for some of the species appear to predate the arrival of humans by tens of thousands of years.[11] Can we forgive such a gap in time by arguing that the fossil record is incomplete?

When it comes down to it, we still don't know which of the two scenarios caused the extinctions. Elsewhere in the world – such as with the loss of mammoths and giant deer in Europe and Asia – it was almost certainly the combined effect of humans hunting populations that had already been diminished by climate change that drove them into oblivion. For Australia, people's preferred argument seems to hinge on whether they accept the absence of evidence that humans hunted many of the species as evidence that they didn't, alongside data that show the climate was harsh; or if they prefer the appeal of mathematical models that imply that it would have been easy for humans to kill them off if they did co-exist. The jury is very much still out.

Whatever the cause, whenever I'm out in the vast grasslands of Australia, it's impossible not to think of all the giants that were there just a few tens of thousands of years ago, as well as the big-cat-sized predators. Not only would it have been a sight to see, but – from possible equivalents of pandas and sloths to rhinos and lions – it's interesting to note that some of Australia's missing marsupials may have filled similar niches to other mammals elsewhere in the world. This is a recurring theme across the country's mammals.

10

Copycats and Cover Versions

When the same or similar features evolve independently in different groups of living things, scientists call this 'convergent evolution'. It can happen at a molecular level – in the way that the proteins in platypus venom have evolved to be so similar to those found in spiders and anemones – or at the level of major anatomical features. For example, complex eyes with lenses have appeared separately in vertebrates, cephalopod molluscs, box jellyfish, bristle worms, velvet worms and arthropods. None of their shared ancestors had such eyes, yet each group acquired them independently. The evolutionary benefits of being able to see the world are so enormous that eyes keep popping up on various branches of the tree of life.

Aside from the giants in the last chapter, there are some fantastic examples of convergences between marsupials and placental mammals. For example, the striped carnivorous marsupials thylacines evolved near-identical body, head and skull shapes to members of the dog family, which are placentals with similar habits. Likewise, Tasmanian devils (marsupials) and hyenas (placentals) have evolved similar skulls and teeth for crushing bone in order to eat animal carcasses. Wombats (marsupials) and marmots (placentals) both have similar adaptations for burrowing and ever-growing teeth for eating tough alpine plants. Possums (marsupials) and New World monkeys (placentals) both have fingerlike appendages on the end of their prehensile tails, which act like a fifth limb for use in climbing trees. The various kinds of gliding possums (marsupials) and flying squirrels (placentals) have both evolved parachutes of skin between their limbs to glide between trees. Marsupial moles and both African golden moles and the talpid moles of Europe, Asia and North America (placentals) have evolved tubular bodies with silky fur, reduced eyes and massive digging forelimbs for underground living.* Similar features have appeared in each of these pairings as similar adaptive solutions to similar evolutionary problems, even though one member of the pair is a marsupial and the other is a placental. The last common ancestor to these two groups lived around 160 million years ago and wouldn't have had any of these specialist adaptations – they evolved independently and more recently in each group.

My very favourite example of convergent evolution is between the striped possums and the aye-aye, the nocturnal wood-pecking lemur of Madagascar. Few animals have robbed me of more sleep than striped possums. Not because they are noisy, but because I couldn't find any, having travelled for days across tropical Queensland in order to do so. These long-tailed, strikingly black and white,† squirrel-sized marsupials are strictly nocturnal, so each night for a week I would try to sleep for ninety minutes, then get up and search the rainforest around my tent for thirty minutes, then repeat. By the last

---

* One of the most commonly used examples of convergence is between the sabre-toothed cats and the extinct South American relatives of marsupials with massive sabre teeth, called *Thylacosmilus*. This name, which means 'pouched knife', conveys a link to the most famous sabre-toothed cat, *Smilodon* – meaning 'knife-tooth'. However, a fascinating paper in 2020 concluded that *Thylacosmilus* could not have used its canines to kill like *Smilodon*, and in fact fed like no other known animal. One possibility was that it used its massive canines to tear open animal stomachs, and then used its tongue to suction-pull the guts out, rather like how a walrus uses its tongue to shuck clamshells (Janis et al., 2020).

† Their strongly contrasting markings are believed to be warning colours against the fact that they give off a noxious chemical when attacked. Very few mammals have warning colours (like bees, wasps or poison dart frogs do).

night I was dead on my feet and devastated by my lack of success. Then, at 5.30 a.m. on the final morning – the very last opportunity before the sun rose on the day I had to leave – a striped possum popped up in a tree at head height and within touching distance. It couldn't have been a better encounter, and I was able to see with my own eyes the adaptations that they share with aye-ayes, through convergent evolution. Not only have both species evolved identical-looking teeth for biting holes in trees, but they also possess an identical-looking elongated finger for gouging beetle grubs out of those holes (although the aye-aye's third finger is the longest, whereas it's the fourth in striped possums).[1] Striped possums also have the largest brain relative to body size of any marsupial, again showing a similarity with primates.

Some of the marsupial/placental species pairs above are the key go-to examples for explaining convergent evolution. Tellingly, however, the way people typically talk about these convergences is directional. They tend to say striped possums are 'marsupial versions' of aye-ayes, that thylacines are 'marsupial versions' of wolves, and that Tasmanian devils are 'marsupial versions' of hyenas. It is never the other way around. I've never heard anyone say that hyenas are an African or Asian version of a Tasmanian devil. They say possums are like monkeys; not that monkeys are like possums.

This implies a hierarchy. It's another subtle and unconscious hint that placental mammals are superior to marsupials. Much as in music, the cover version is never considered to be as good as the original. But this is not how convergence works.

*

On the planet right now, there are exactly four museums that contain the preserved bodies of adult thylacines. Not taxidermy or unmounted skins, not skulls or skeletons, but jars of preservatives holding the thylacines as if in suspended animation.

They are the National Museum of Australia (a skinned male), the Swedish Naturhistoriska riksmuseet (a female considered the finest specimen of its kind), the Oxford University Museum of Natural History (which has two – one partially skinned male and one headless female), and the Grant Museum of Zoology at University College London (which has been dissected into four pieces), where I worked as the museum manager. There are around 800 thylacine specimens in total in the world's museums, but only five are fluid-preserved adults.[2] I consider them to be the most palpable connections we have with any extinct species; the eternal destructive process condensed into five physical objects.

In my fourteen years at the Grant Museum, I briefly removed the thylacine from its jar only once. It stands as the most profound experience of my museum career. I have handled many thousands of dried, skinned and skeletal mammal specimens in my life; they do not feel like animals – they feel like inert *things*. But holding in my gloved hands the visceral weight of a dead thylacine – the species that most embodies modern extinctions – was altogether different. The thylacine is an animal that has gnawed its way into my psyche – its social and natural histories have long been core passions of mine. Whenever I am in Tasmania, every firebreak, clearing or path junction I pass, I cannot stop myself from turning my head to see if one is there. It is subconscious and illogical.

The last known thylacine died of exposure, locked out of the indoor part of its zoo enclosure on a cold Tasmanian late winter's night on 7 September 1936. While there is little doubt that they survived in the wild for several decades after that, I do not believe (as many do) that they still cling on in the wilder parts of Tasmania.* Nonetheless, my brain cannot stop

---

* There have been very many reports of sightings, of which none of the recent ones are particularly plausible; however, one of the most credible post-1936 accounts involved David Fleay himself. In 1945, in order to try to catch a thylacine for captive breeding, he began a months-long expedition to southwest Tasmania (part-funded

my eyes from looking for them. The few moments it took to lift that thylacine carcass out of its specimen jar were, for me, electric. The way the weight shifted as its limbs moved as I turned it, the inconsistent density of a large carcass, its corporeal flexibility, the feel of flesh beneath the skin – it was more real than any specimen I have worked with. It wiped away the more than 100 years since it was a living, breathing thylacine. I know how this sounds, but I felt connected. Museum specimens are not just scientific data – they are artefacts of human history and our relationship with the world.

Powerful zoological ley lines intersect in this specimen. Not only is it a thylacine – iconic in its own right – and not only is it one of only five preserved adults in existence, but it belonged to Thomas Henry Huxley, one of the most influential scientists ever. Huxley was largely responsible for Darwin's evolutionary theories being accepted by the Victorian scientific community. It was he who stood up to Richard Owen to fight Darwin's corner.

If you put a thylacine and a dog or dingo side by side, the overall similarities are striking – you would have no reason to think they were separated by 160 million years of evolution. The only glaring difference is that one has stripes. When the thylacine's genome (the details of its full genetic make-up) was published in 2017, the authors investigated the underlying nature of the convergence. They found that despite the fact that anatomically, thylacine skulls were far more like the skulls of members of the dog family (canids) than they were to any

---

by Keith Murdoch, father of Rupert). His daughter Rosemary describes finding the aftermath of what they are convinced was a near-miss catching of a thylacine in a leg-hold trap at Poverty Plains: 'We came on the scene . . . and found plenty of definite footprints in the soft wet ground, some hair was adhering to the traps which was gathered and when analysed later at the Tasmanian Museum was proven beyond any reasonable doubt to be hair from a Thylacine.' (Fleay-Thomson, 2007). Unfortunately, that was just ten days before the scheduled end of the expedition. Fleay made plans to return the next year, but they were scuppered by the travails of getting the live platypuses to New York.

of their marsupial relatives, these animals had not undergone similar genetic changes in their evolutionary histories.[3] This means that thylacines and canids evolved close physical similarities, but these were not brought about by evolving similar genes.

Darwin himself cited the similarity of thylacine and dog jaws as an exemplar of convergent evolution in the sixth edition of his groundbreaking book, *On the Origin of Species* – inarguably the most important book in the history of science.*[4] For those of us who have become somewhat obsessed with the thylacine's story, a book written in 2000 by Robert Paddle may also hold a position of significant influence. *The Last Tasmanian Tiger* is the most complete account of the history of the thylacine and why it is so important. In it, Paddle provides a fascinating insight into the role the species played – before its extinction – in the battle over evolution.

Two different groups attempted to use the same basic facts – the shared features between canids and thylacines – as evidence for two diametrically opposing views about the origins of species. To set the scene, we need to explore how certain descriptions of the thylacine were unreasonably critical – it is a familiar story of attempts to paint a marsupial as inferior with questionable evidence. Paddle quotes accounts that suggest that the thylacine's limbs were out of proportion to its body and its head was overly large, giving the impression that it was 'badly formed and ungainly and therefore very primitive' (despite its tone, that particular quote is actually from Michael Sharland, who was one of the strongest

---

* Incidentally, despite the fact that the similarities between thylacines and wolves is one of the most ubiquitous examples of convergent evolution, a study in 2021 compared the skulls of thylacines and other carnivorous mammals in great detail, and found that rather than wolves, African jackals and South American 'foxes' (such as the maned wolf and Pampas fox) were actually the closest comparators to thylacines. This is important ecological information, as all these canid species hunt prey less than half their own size – giving us insight into the way thylacines are likely to have lived. (Rovinsky et al., 2021)

advocates for legally protecting the thylacine as it slid into oblivion). These descriptions place value judgments on subjective aesthetic appearances of an animal. As Paddle puts it:

> At times the constraining power of placental chauvinism upon scientific thinking almost beggars belief! Were [thylacines] assumed to be a placental carnivore, the large head in proportion to the body would be interpreted as an obvious sign of an animal of high intelligence. However as [it] is known to be a marsupial, then the large head is a sign of a primitive and ungainly physiology. Truly, if you are born a marsupial you just can't win![5]

It was such anti-marsupial biases – or 'placental chauvinism', as Paddle puts it – that formed the basis for an evolutionary argument around the thylacine. On one side there were people like Thomas Henry Huxley, who had dissected the thylacine that is now in the Grant Museum. Huxley espoused a particular view of evolution that assumes that it is directional and progressive. You know now that evolution should not be viewed like this – no one species or group is more or less evolved than any other. Huxley did not see things that way, however, which was not particularly unusual for naturalists of his day. He and others with a progressionist view of evolution were adamant that placentals were a more advanced kind of mammal than marsupials – a higher step on the evolutionary ladder. The thylacine was a case in point, they argued – it had evolved predatory adaptations that made it look like a dog, but they argued it was inferior to it in the kinds of ways outlined above (and also just because it was a marsupial). Evolved . . . but not quite evolved enough.

On the other side, some creationists used the same underlying zoological observations of the animals to reach the opposite conclusion: that there was no evolution. They would look at the similarities between thylacines and canids and claim that they were evidence for the perfection of creation. A creator had designed the ideal features required for a four-legged predator in the kind of habitat that thylacines and wolves inhabited and given those same features to both creatures.

Most biologists today would say that one aspect of the creationists' view is correct: that the two groups – thylacines and canids, marsupials and placentals, or one animal and any other – are essentially equals. None is hierarchically superior to another. However, it was evolution, not creation, that made them that way – albeit not in a stepwise progression 'up' a tree of increasing quality.

Not everyone who believed in the divine creation of species had an egalitarian view of the world. We have already heard that Richard Owen made judgmental statements that denigrated marsupials – he clearly viewed them as substandard mammals. Twenty-five years before Darwin first published *On the Origin of Species*, Owen, in a list of apparent marsupial deficiencies, suggests that their vocal range also implied inferiority:

> Another character . . . is the want of a power of uttering vocalized sounds. When irritated [marsupials] emit a wheezing or snarling guttural sound; . . . The Thylacinus or large Dog-faced Opossum, [George Harris] observes, utters 'a short guttural cry, and appears exceedingly inactive and stupid, having, like the owl, an almost constant motion with the nictitating membrane of the eye.' The Wombat, when irritated, emits a loud hiss which forcibly reminds one of that of the Serpent.[6]

Invocation of 'the Serpent' is the lowest blow a religious man like Owen could make. And as Paddle points out, it is deeply unfair to quote Harris's account of the thylacine as a means to convey information about their intelligence. Harris was the deputy surveyor-general of New South Wales, and he wasn't talking about his experience of how thylacines live their lives, he was specifically describing one thylacine in 1806 – the first to be caught and illustrated by a European – and the effects of its wounds while it spent a few hours dying in a cage trap.[7] It wasn't 'inactive and stupid' because thylacines are unintelligent, it was 'inactive and stupid' because its life was painfully eking away. Owen knew that but decided to use Harris's account to paint all thylacines as dim and second rate – and in a particularly weak argument that aligns an animal's perceived quality with the noises it makes.

Unfortunately, the habit of casually implying that thylacines are inferior versions of mammals does continue to slip into the writings of modern scientists who hold the animal in high regard. In 2018, a number of expert thylacinologists for whom I have the greatest respect published a fascinating paper that – for the first time – outlined the way that young thylacine pups grew in the pouch, by CT-scanning the eleven known fluid-preserved infant specimens surviving in museums. The point they were trying to make was that the marsupial reproductive strategy had placed a constraint on the evolution of thylacine limbs. Thylacine babies need to have flexible limbs in order to climb into the pouch, and this is what the researchers saw in the CT scans of the developing pouch young. Because of this need, they postulated, the convergence seen in the heads and bodies of thylacines and canids did not extend to thylacines evolving less mobile limbs. Reduced flexibility in the legs is an adaptation that we see in members of the dog family that is associated with running and pouncing. The way I see it, this means that in thylacine evolution, the crawl to the pouch is more important than a stiffened wrist when attacking prey. However, I winced when I read this line in the article: 'Despite [thylacines] sharing striking similarities with canids, they failed to evolve many of the specialized anatomical features that characterize carnivorous placental mammals.'[8]

It is so easy to write a sentence like that and not notice

what it says. The phrase 'failed to evolve' implies that if only it had the chance, a species would ideally evolve to be like a placental mammal. I hesitate to use the phrase 'placental chauvinism' here, as Robert Paddle himself was one of the authors of this paper, but I implore zoologists and journal editors everywhere to be careful about slipping this kind of language into publications. Marsupials have not 'failed to evolve' to be more like placentals. Again, placentals like us are not the pinnacle of evolution, and it isn't reasonable to suggest that in the struggle for existence, animals are striving to be more like us. So systemic is the bias towards the supposed superiority of placental mammals that even staunch advocates of marsupials occasionally slip into implying their inferiority. Even if one were to argue that the way dogs run is more efficient, the fact that thylacines need their limbs to be configured in a different way is not an evolutionary failure.

*

Thylacines have been given several names that compare them to placental mammals, but as Carol Freeman points out in *Paper Tiger,* her history of how thylacines have been depicted in art, these names have not been consistently used over time.[9]

When Harris formally described the species in 1808, he gave them the name *Didelphis cynocephala*, meaning 'dog-headed opossum'.[10] Descriptions of the thylacine as being dog-like then disappeared from the literature for the following thirty years, and illustrations of them didn't depict them as being particularly dog-like until near their extinction. Instead, we see thylacines compared to other carnivores. They have been called various combinations of the words marsupial, Tasmanian or zebra, with wolf, hyena and tiger. Unlike the 'tiger cat' moniker for spotted-tailed quolls discussed previously, it is at least possible to see the similarities here, as thylacines are stripy and have a similar body shape to wolves or hyenas. It's worth noting, however, that wolves and hyenas are also species that have been demonised by humans and viciously persecuted as a result, unlike dogs. The sheep-farming lobby in Tasmania tirelessly (and inaccurately) vilified thylacines as livestock-killers. While wolves and hyenas have been driven to extinction across much of their ranges by farmers seeking to protect their animals, thylacines were completely wiped from the face of the Earth. I think it is no coincidence that they were given common names like 'marsupial hyena' and 'zebra wolf', which allied them with these placental mammals that humans considered to be enemies – it made it easier to justify the bounty system that caused their disappearance. We know that a similar tactic was explicitly pursued with a different unpopular predator on the Australian mainland: one hunting organisation fined their members if they referred to dingoes (which were also considered a pest on livestock) as anything other than 'Australian foxes', in order to increase the desire to hunt them.[11]

Pictures were also employed to influence the public's opinion of thylacines. There are plenty of nineteenth-century illustrations of thylacines that were clearly derived from sympathetic original depictions made from life (particularly the pair that arrived at London Zoo in 1850 and appeared in Gould's *The Mammals of Australia*) to which wolf-like features were added. These thylacines were given shaggy fur, menacing expressions, muscular necks and prowling or pouncing stances. Captions accompanied these foreboding visuals that claimed that they were as bloodthirsty for sheep as wolves.[12] Harry Burrell, our hero of platypus biology, staged a widely circulated photograph appearing to show a thylacine walking off with a chicken, but it was in fact a taxidermy specimen with a hen tied to its mouth.

Despite the extraordinary pressure put on them by habitat loss, the government-sponsored bounties and the privately organised ones that preceded them, thylacines were said to be at fault for their own extinction. Just as happened with other

This thylacine has been drawn to look particularly fierce and wolf-like, complete with a mountainous backdrop, which accentuates the 'wild' and 'savage' depiction of the animal.[13] This 1865 illustration by Robert Kretschmer was well circulated, appearing in a number of nineteenth-century popular natural history books.[14]

marsupials, once they were gone, people attempted to wash their hands with the blood of the vanished thylacines by claiming that they were too stupid, too unadaptable, or otherwise too poor a version of a mammal to survive.[15]

Names, images and written descriptions are all important factors in how the public considers animals, and as we are about to explore, so is how their physical remains are presented.

11

They're Stuffed

Natural history museums are generally considered to be relatively objective, apolitical, scientific places. Many people tend to trust what they see in museums far more than other sources of information. However, we should remember that our museums are a direct product of empire, historically based on colonial ideals. As such, they are inherently subjective and political. Collecting was part of the act of colonisation and facilitated the exploitation of colonial resources – animal, vegetable or mineral. Unsurprisingly, as someone who works in one, I think natural history museums today are wonderful. Their collections are vital in enabling us to answer global challenges like biodiversity loss and climate change, and the key purpose of their galleries and programmes is to inspire an appreciation for the natural world (among many other positive things). However, remembering their historical roots allows us to try to unpick precisely how accurately museums do represent nature, and people.

If museums were truly unbiased reflections of global biodiversity, we would expect to find an even spread of where in the world their specimens come from (and for 80 per cent of their collections to be arthropods – as insects, arachnids, crustaceans and their relatives make up that percentage of described species today – rather than mammals and dinosaurs being so massively over-represented). Instead, however, museum collections are geographically biased in ways that reflect their countries' colonial histories. For example, the UK has relatively few Chinese or Russian specimens, but a mountain of Indian, Australian and African ones. To illustrate this, I searched how many specimens from certain countries are listed on London's Natural History Museum's online database. Over 100,000 came from Australia, and over 87,000 came from India, Pakistan and Nepal combined. By contrast, only 36,000 originate from China and 12,500 from Russia. In reality, New Zealand is significantly dwarfed by both of the latter, yet it has 27,000 specimens.* French, Belgian, Dutch, Portuguese and German collections are similarly biased to their former colonies. I was recently lucky enough to explore the stores of Naturalis Biodiversity Center, the Netherlands' national collection, and my jaw dropped at the sheer number of rarities and type specimens (the specimens on which the original species descriptions were based) of marsupials and echidnas from New Guinea and other parts of Indonesia – formerly in the 'Dutch East Indies'.

Echidna and platypus specimens are not museum rarities. In countries with imperial links to the places in which these animals are found, they are almost universal features of museum collections. (Part of what we in the business call 'museum bingo' – where you can seek out the same species in nearly every natural history museum in a given country.[1]) In the UK, platypuses and short-beaked echidnas are universal, due to our colonial relationship with Australia. In Germany and the Netherlands, long-beaked echidnas are universal, due to those countries, colonial relationship with New Guinea.

*

---

* My colleagues at the museum would want me to point out that this is a rather crude way of making this point, as their collection houses over 80 million specimens, of which 'only' around 4.5 million are so far online.

It's not just the proportional make-up of museum collections that are unnaturally skewed. Despite being scientific institutions dedicated to providing a window on the natural world, the specimens exhibited in natural history museums are often inaccurate. I have reached the conclusion that no group is more consistently depicted wrongly than Australian mammals, and of these none more so than the echidna, the single most badly stuffed species in the world. Not only are the vast majority of museum specimens the wrong shape, but they often also have their feet pointing in the wrong direction. In short, they do not look like echidnas.

As you now know, in real life, echidna hind feet point backwards. Unfortunately, however, this is not reflected in most echidna specimens in museums, whose feet have been twisted all the way around to point forwards.

Echidnas do not, as far as we know, twist their own feet around after death. From conversations with museum visitors, it turns out that a reasonable number of people are under the impression that the specimens in museums are simply dead animals that have been somehow preserved whole as they lay, and then just placed on the shelf.* In fact, taxidermy is a highly skilled art form that requires a great deal of work to transform a carcass into what we see on display. It involves the pursuance of an illusion in which the viewer is given the opportunity to forget that the animal is dead. It has to look alive.

As soon as an animal dies, it is a race against time to get to work before decomposition begins and the skin 'slips',

---

* Technically, in the case of freeze-dried specimens, this could be the case. Freeze-drying is a process whereby all the moisture is removed from a whole carcass, to leave it looking pretty much alive, but rock-solid (and eyeless). There is a limit to how large a specimen can be for this process to be successful, and even an animal the size of an echidna is probably too big. I was once doing an assessment for a university collection that had been neglected and unused for some years, and opened the lid on a black bin to find a whole freeze-dried chimpanzee inside. Chimps are far too big for freeze-drying, and they had clearly failed to remove the water content from the tissues in its core, so it was slowly rotting. It was grim.

causing the fur or feathers to fall out. The taxidermist must carefully skin the carcass in one piece, making the minimum number of cuts possible in the least visible places. They will often leave the bones of the hands and feet inside the skin due to the difficulty of extracting them all neatly, and also because that allows the finished article to look more realistic. There is relatively little soft tissue around these bones, which reduces the risk of rotting. The taxidermist then tans the flayed skin with some rather unpleasant chemicals to halt the decomposition process. That's the easy part.

The real skill is in reforming the empty skin into the appearance of a real, living animal. To do this, the overall shape of all that went beneath the skin needs to be recreated. Despite the common use of the word 'stuffed' for taxidermy specimens, it is absolutely not the case that they are created by sewing the skin back up and stuffing a load of cotton wool inside (that would be what we call a study skin or cabinet skin – just a sausage-shaped tube for easy and stable storage, for use in research rather than display). What is required is an internal 'form' for the taxidermy mount, recreating the animal's overall appearance while also acting as the support structure to hold it up.

Today, premade forms for more common taxidermy animals can be purchased online in a variety of poses, and the taxidermist then just has to slip the skin over it and sew it up (I'm making that sound a lot easier than it is). Alternatively, as it was historically, they can make their own. This could be carved whole from wood into the rough shape of the animal or built up by binding a padding material (for example, textile or wood wool) around an internal scaffold of wire, welded metal rods or wooden batons. Taxidermists I know ensure the form is accurate by measuring or drawing around all the internal packages of muscle on the skeleton after they skin the carcass. Often, the de-fleshed and cleaned skull will be put to use back inside the skin, to provide the most accurate shape as well as ensuring that the teeth are correct (and for anything with bony horns or antlers, at least the top part of the skull is always required). Otherwise, teeth can be formed out of plaster, wood, resin or plastic, which would also be used for moulding realistic-looking gums and tongues before painting.

In all this, the taxidermist – or their client – must decide what pose to put the animal in. Is it sitting, squatting or standing? Which way is it looking? Is it flat on the ground, on a rock or a branch? Is it interacting with any other animals or objects? What is its facial expression? These choices can all be used to exhibit some known behaviour, like a predator crouching in ambush, a squirrel scurrying up a tree or a bird in its specific nest.

Glass or plastic eyes need to be added, too. Different species have very different eyes. Although few people could describe what the colour or shape of the iris of any given animal's eyes looks like, it is surprising how noticeably odd a specimen looks if the wrong eyes are used. Again, today these can be mail-ordered for a wide range of animals, but otherwise they would need to be handmade.

The word taxidermy comes from the Greek for 'drawing or pulling the skin', and that's what happens next. Far from 'stuffing' it, the taxidermist must precisely shape the skin around the internal form that they have made, and then sew it up in a way that the stitches are invisible. A lump or bump out of place – or an anatomically unlikely wrinkle (or not enough wrinkles) – will destroy the illusion. Naked areas of skin around the faces, hands and feet are then painted to reinvigorate their perfectly matched colour.

Needless to say, there are a lot of opportunities for even the most skilful of taxidermists to get something wrong. Historically, taxidermists in Europe or America often had little source material to guide the transformation of a flat animal skin sent from the outposts of empire into a taxidermy mount for museum displays. If they were very lucky, they were sent a sketch of the animal along with the skin, occasionally they would have been given written descriptions, but very often it was just a skin in a barrel. They had to use their assumptions of what shape the species was, based on their (often extensive) experience of anatomy. They assumed – reasonably – that animals' feet pointed in the same direction as their heads. But that is not the case for echidnas.

In order to force the feet on an echidna specimen into these forward-facing, incorrect poses, the skin on the ankles would often rip as they twisted it round. But still they didn't realise.

These mistakes also extended to mounted skeletons, which had their leg bones twisted out of position to fit the assumption that feet face forward. And it's not just the legs on taxidermy echidna specimens: the entire body shape is typically wrong, with some specimens being mounted in an upright horse-like pose that may have been inspired by assuming they look like European hedgehogs, while others are shown completely flat and frog-shaped with their bellies on the floor and their limbs sticking out spread-eagled (it's hard to imagine how an animal would move in that pose), and yet others are like rugby balls.

Sometimes, they have their snouts reshaped in plaster and pointed in different directions (up, down or forwards). I saw one short-beaked echidna in Leiden that had been given a snout in the shape of a downward-curving carrot, presumably confused with a long-beaked echidna, which do have snouts shaped like that.

All in all, few specimens are actually echidna-shaped. We can tell that confusion must have been widespread by noting that no two taxidermy echidnas look the same. One of our taxidermy echidnas at the University Museum of Zoology in Cambridge has its feet pointing in the correct direction; however, it is the fattest specimen I have ever seen. It's one of those that has been mounted with its belly flat on the ground, but it's been overstuffed to such an extent that its four legs have been subsumed into its bulk, leaving just the claws of its digits to stick out, like a novelty balloon. It has no neck, and its beak also points straight out of this pillow-like (but spiky) specimen. It reminds me of a spiny loaf of bread with claws. I've never seen an animal – of any species – adopt anything like this pose.

Wrong-footed and wrongly shaped echidnas are even common in Australian museums. Ayesha Keshani is a researcher in visual cultures at Goldsmiths, University of London, who I heard give a paper on the Muzium Sarawak in Malaysian Borneo at a conference on decolonising natural history museums. When she explained how many of their specimens had been collected in Sarawak, shipped to London to be mounted and then sent back to Borneo for display – thereby imposing Western representations of Bornean animals on a museum for Bornean people – I wondered if the same thing had happened with echidnas. How many of them, after having been caught in Australia, were sent to Europe to be mounted as taxidermy and then sent back to museums in Australia?

In the past, taxidermists often based their work on drawings, but those drawings often depicted other incorrect taxidermy echidnas (all but one of my own collection of historic echidna illustrations has its feet pointing in the wrong direction). Or, one taxidermist would copy another, who had got it wrong. It was a constant cycle of copying incorrect representations, each with the opportunity to add their own errors along the way. I have discussed this with Alison Douglas, museum preparator and taxidermist at the Queensland Museum (it is relatively rare for museums to have taxidermists on staff these days – not a single English museum employs one), who told me that ironically she thinks echidnas are actually one of the easiest species to prepare as taxidermy, as their general shape is so simple and cushion-like. Unlike many of her early predecessors, Alison has the benefit of skinning her specimens herself, and growing up in Australia has afforded her plenty of opportunities to see living, moving echidnas (not to mention access to Google images) to inform how she shapes her creations – which, I think, are perfect.

Similarly, platypus taxidermy is regularly misshapen. We commonly find 'fattypus', 'flattypus' and 'platysausage' specimens in museums. As with their echidna cousins, the people preparing the skins in Europe didn't know what shape they were when they were alive, so they had to rely on their knowledge of other animals' anatomy, which are not a good match. Taxidermy platypuses are occasionally given golden eyes, when in truth they should be black. This is also seen in some historic illustrations and written descriptions of platypuses. It's hard to know where the error first came from (taxidermist, writer or illustrator), but it's easy to imagine that a taxidermist popped the wrong eyes in their specimen, and others copied from there.

A few years ago, I was flummoxed when I encountered a taxidermy echidna with electric blue eyes at the Manchester Museum. Where could that error have come from? Manchester's mystery echidna popped back into my mind when rereading the original English description of the species, and I think I now know the answer.

In 1792, George Shaw's otherwise perfectly reasonable description of the species includes this passage:

> The snout is long and tubular, and perfectly resembles in structure that of the . . . great ant-eater . . . The nostrils are small, and seated near the extremity of the snout. The eyes are very small, and black, with a pale-blue iris.[2]

The question then becomes why Shaw thought echidnas had blue eyes. In any case, it seems likely that the echidna in Manchester was modified to fit Shaw's original description.[3] Curiously, the illustration that accompanies the text has correctly coloured black eyes. And actually, looking at it again now, that engraving appears rather flat-bellied, shapeless, neck-less and with its hands and feet sticking straight out with no legs in between. Perhaps Shaw's paper was inspiration for our overstuffed specimen in Cambridge.

*

Sometimes, taxidermists have everything they ought to need to mount a specimen correctly, but still get it wrong. The introduction to this book started with the story of the platypus that Governor John Hunter sent to the Literary and Philosophical Society in Newcastle upon Tyne. As I mentioned, in the same barrel was the first wombat to make it to Europe. And this is its story.

In February 1797, after a troublesome journey from Calcutta (Kolkata) in India to the colony at Port Jackson – the first European settlement in Australia (which grew into modern-day Sydney) – the ship *Sydney Cove* floundered as it reached its destination. The captain wrecked it in a sheltered bay of what is now called Preservation Island, one of the Furneaux Group of islands in the Bass Strait (the water between Tasmania and mainland Australia). The whole crew and most of the provisions they intended to sell made it to dry land (the captain stashed the rum on nearby Rum Island to keep it away from the crew). Eighteen men sailed for the Australian mainland in one of the ship's boats and set out to walk to the colony to seek help. In May, just three of them reached Sydney. Those who remained on the island explored while they waited, and on the neighbouring islands they found wombats.* It strikes me that three of Europe's most significant early encounters with marsupials only came about through shipwrecks: first

---

* One of these islands is now called Badger Island – 'badger' being a name used by early settlers for wombats (and you still hear it in Tasmania today), although they cannot be found on Badger Island now. Flinders is the only one of the islands where the Bass Strait wombat – a unique subspecies – survives.

Pelsaert and the tammar wallabies, then Cook and the kangaroos, then the *Sydney Cove* and the wombats.

Over the course of ten months, Governor Hunter sent a
schooner called the *Francis* on three trips to rescue the crew
and cargo from Preservation Island. On her third trip, with
Matthew Flinders (the man who would later confirm that
Tasmania was an island; circumnavigate Australia; and who
was most responsible for the country becoming known as
'Australia') on board, they took a live wombat back for
Hunter.

It arrived in March 1798 in a weak state and died six weeks
later. Hunter sent it – along with the platypus – in the barrel
to Joseph Banks, with a skin and skull of another wombat.
Unlike many specimens that would come to Europe in this
manner, Hunter also sent clear drawings of the animal and a
detailed written description. He did a good job of outlining
its anatomy clearly, and also followed the habit of comparing
a new species to known placentals. Hunter also described the
pouch (what they call a 'false belly'), so Thomas Bewick
recognised it as a relative of the opossums. Three pages of
Hunter's words and an illustration are reproduced in Bewick's
widely read *A General History of Quadrupeds*, published in
1800:

> It is about the size of a Badger, a species of which we
> supposed it to be, from its dexterity in burrowing in the
> earth, by means of its fore paws; but on watching its general
> motions, it appeared to have much of the habits and manner
> of a Bear.[4]

After this groundbreaking wombat arrived in Newcastle, it
does not appear to have received the attention it deserved
(perhaps they were too distracted by the platypus), and its
preserved carcass was allowed to dry out after the barrel burst
over that poor woman's head. This is how it was left for thirty

Governor John Hunter's drawing of the wombat that he sent to England with the platypus specimen, reproduced by Thomas Bewick in 1800.[5]

years, until it was prepared for mounting by local taxidermist Richard Wingate, one of Bewick's close friends. Despite all the first-hand information about how wombats look that Hunter had provided, and that Bewick had reproduced, Wingate decided to trust his anatomical instincts, and interpret the carcass in his own way.

He noticed the thickened skin of its haunches (wombats really do have notable bottoms) and concluded that they were calluses formed by the wombats sitting upright on their backsides, and that this was how they generally carried themselves, rather than walking on four legs as Hunter had described and drawn.[6] And so, he posed the animal sitting up on its bottom, like a kangaroo, with its arms held out somewhat *Tyrannosaurus*-like. To this day, 'the governor's wombat' can still be seen on display at the Great North Museum: Hancock in Newcastle, inaccurately standing on two legs. In homage to Europe's first wombat, the Tasmanian Museum and Art Gallery has produced a taxidermy wombat in the same pose for its visitors to admire.

*

Aside from generic inaccuracies, there are other notable pecu-
liarities in the way Australian mammals appear in museums,
particularly with respect to the thylacines (which, like echidnas,
are also regularly misshapen by taxidermists, but that's not
what I'm talking about here).

In order to tell this story, I need to make a brief diversion
into the history of penises in museums. Many people know
that Queen Victoria insisted that the genitalia on statues in
London's Victoria and Albert Museum be covered by plaster
fig leaves whenever a female member of the royal family called
in (when the museum is named after you, you are allowed
these little riders). Prior to each visit, the curators would run
around with their stepladders, carefully concealing the crotches
on artworks such as their replica of Michelangelo's *David*.

However, most people don't know that natural history
museums also modified their specimens to cater for the sensi-
tivities of Victorian prudes. The males of most species of
mammal have bones in their penises. The bone is called a
baculum (plural, bacula), meaning 'little staff or stick' in Latin,
and is found in primates, rodents, shrews, hedgehogs, moles,
carnivorans (cats, dogs, bears, seals, weasels, raccoons,
mongooses and their relatives) and bats. If you can remember
that the taxonomic term for bats is Chiroptera; *and* you can
forgive the fact that 'Insectivora' is an outdated taxonomic
name for hedgehogs, moles and shrews; *and* that carnivoran
isn't spelt with a K, this is a handy mnemonic should you want
to be able to rattle off which beasts are bestowed with bacula:

> P rimates
> R odentia
> I nsectivora
> C hiroptera
> K (C)arnivora*

---

* Irritatingly, the American pika – a relative of rabbits and hares – was recently

These are all large groups of mammals. In fact, if you were to add up the number of species in each group, it would amount to around 80 per cent of all mammals (although there are exceptions in each group which don't have bacula – humans, for example, are primates without penis bones, as you probably know). However, despite the fact that most species of mammal have a penis bone in life, and most mammals on display in museums are male (another example of unscientific bias in museums),[7] you would be hard pushed to find any skeletons with their baculum on display in a typical museum gallery.

It's true to say that because the baculum is tiny in the smaller species, and as it doesn't articulate with any other bones (it just floats in soft tissue), it is extremely easy to lose when de-fleshing a skeleton. This may explain its absence in many instances. However, that doesn't excuse why they aren't attached to larger skeletons – a walrus's penis bone, for example, can reach 60 centimetres (2 feet) long. It's hard to accidentally miss that.

No, in such cases – with disregard to their institutions' scientific integrity – the preparators and curators have deliberately removed the penis bones to spare visitors' blushes and sniggers. In similar ways, in taxidermy specimens most animals' genitals are quietly hidden from view.

But not so the thylacine. For some reason, the trend is to see male taxidermy thylacines with extremely prominent scrotums. And I mean *really* prominent. It's hard to understand exactly how or why this became the fashion. One might suggest that it is intended to highlight the difference in the way that marsupial genitals are arranged: the penis is behind the balls* (and is often also forked, to match the forked vaginas in the

---

discovered to have a penis bone. Their group name is Lagomorpha, but 'PRICKL' just doesn't have the same ring to it. Also, my American editor, Joe Calamia, tells me that 'prick' is a bit of a Britishism, and isn't widely used as a slang term for penis in the USA, so apologies to American readers for whom this mnemonic is meaningless.

* This is also true of most rabbits and hares.

females). However, this explanation doesn't really fit, for two reasons. First, while I do think that scrotums feature more commonly on marsupial taxidermy than on placental taxidermy in general, the trend seems far more pronounced in thylacines. And second, the penis isn't typically included in the taxidermy anyway, so you can't see how it is positioned in relation to the scrotum. Live marsupial penises are usually kept safely tucked away inside the cloaca until they are needed, at which point they pop out. As far as I am aware, only one of the world's taxidermy thylacines has an erect penis (which is extremely rare in taxidermy of any species – I can't recall ever seeing another species with a taxidermy erection), and that's one of the oldest thylacines to leave Australia (listed as 'pre-1824'), at the Naturalis Biodiversity Center in the Netherlands. Its penis is over 20 centimetres (8 inches) long, but we'll never know whether that is accurate – it seems unlikely. It's no surprise that the specimen is not on public display.

One thing that is unusual about thylacines is that the males had a pouch into which they could retract their scrotum, giving it an extra layer of protection (the male water opossum also has a pouch, which is thought to protect the balls while swimming and also helps to make them more streamlined). Is this why such a big deal is made of thylacine scrotums? No. Because again, this isn't communicated through the taxidermy animals, which just have prominent balls, with no pouch included. In fact, having a scrotal pouch makes thylacines an even *less* likely species to be singled out by taxidermists to highlight their scrotum. Curiously, there are several Indigenous rock art depictions of thylacines on the mainland that also have prominent pendulous balls hanging below the tail. However, it seems unlikely that this is because they were a noticeable feature of the live animals. When we look at the more than 110 photos of captive thylacines, the scrotum is visible in almost none of them. I really don't know why this trend became established.

Another thylacine taxidermy trope is that they are often given a vicious snarl. This reminds us that taxidermy isn't all that it seems: it can convey a political message. In this case, it is suggesting to the viewer that the animal is a rampant sheep-killer that needs to be culled (although we now know this to be untrue – it was a deceit perpetuated by the powerful farming lobby).

In the UK, we see the same snarl given to foxes,* even though it is anatomically impossible for foxes to pull these faces. At the museum I work in, the fox has been mounted with a taxidermy quail in its mouth. I suggest it's intended to hint that foxes are a threat to your game birds and need to be controlled. It's the same story with snarling wolf taxidermy in Europe and America, where this species has been relentlessly persecuted. Society hides its subliminal messages in the poses of preserved animal carcasses.

As I have said, thylacines are arguably the most powerful icons of modern-day extinction. There are so few extinct species for which video footage exists,† which truly brings home the fact that they were living, breathing animals rather than the abstract, lifeless forms we see in museums. Today, curators are putting these specimens to work in order to tell vital stories about the impact of human greed on biodiversity. Look closely

---

* In a curious combination of these tropes, the only thylacine taxidermy in the Czech Republic, at the Národní Muzeum in Prague, has a classic menacing snarl, but the skull inside its head is not that of a thylacine (or a plaster model), but belongs to either a fox or a dog.

† The most famous thylacine footage was filmed in Hobart Zoo in 1933, coincidentally by 'the platypus man', David Fleay, who was bitten on the buttocks by the animal while his head was under his camera's curtain. Fleay's footage – and many well-known photos of the same individual – were believed to depict the last known thylacine. However in 2022 it was discovered to actually be the zoo's penultimate specimen. No known images of the last known thylacine survive. For a long time, Fleay's footage was understood to be the most recent film of a thylacine; however, in May 2020 a new snippet of film showing the same animal was found, filmed in 1935, the year before the animal died.

at many of them and you will spot the cuts to their legs or necks inflicted by the snares that trapped them. In the past, the thylacine was a museum favourite as the specimens could be used to communicate the exoticism of the far-flung corners of empire and the incredible beasts that were found there.

These factors have combined to make thylacines highly desirable objects for museums to put on display at all stages in their European history. This creates a challenge. Putting objects on display causes irreversible damage because of the effect light has on animal skins: it causes them to fade. This is a perpetual dilemma for museum curators: if we display them, we decrease their useful 'lifespan', but people can appreciate them, learn about them and be inspired by their story. On the other hand, if we keep them shielded from the light and covered up in storage, they will last longer, but only those privileged enough to access our storerooms will see them.

The LED lights available to museums today are far kinder on specimens than some of those used in the past (and at least we no longer have oil lamps adding smoke and tar to the issue of fading), but this doesn't stop the deterioration, it just slows it down. Light damage is cumulative: the specimens fade a tiny bit day by day, but over time this adds up.

All of this means that many thylacines in museums are faded beyond recognition. Some no longer bear the stripes that gave the 'Tasmanian tiger' its name. Extinct animal specimens are a finite resource. According to the International Thylacine Specimen Database (which attempts to collect data on every known thylacine worldwide), there are exactly 100 taxidermy thylacines globally.[8] We can never get any more. The very best of them – in my opinion – are on display at the Museum für Naturkunde in Berlin, and in the stores of the South Australian Museum in Adelaide. These two have maintained as near to their living colour as any and are mounted in poses that are not attempting to portray the species as a malicious sheep-killer.

So, what should museums do with faded specimens? Is it still worthwhile having them on display if they aren't accurate representations of the animal when it was alive? For those museums lucky enough to have more than one taxidermy thylacine, should they share the fading evenly and occasionally swap between specimens, or sacrifice one to the light and keep the other safe in store?

When we look closely, we can spot that some thylacines' stripes have been painted on, presumably to reverse the light damage. But as the people who did it didn't document their interventions, we don't know for sure what's under the paint. The painted stripes might increase the display-worthiness of a specimen, but this potentially makes specimens less useful for research as they become less authentic. What's worse is that the chemicals in the paints they used could be causing further damage.

While Bethany Palumbo, former conservator at the Oxford University Museum of Natural History, was assessing their taxidermy thylacine in 2017, she spotted that the stripes had been painted on. There were no records of what paint was used, or when the stripes were added. This presented another ethical dilemma – should she reverse the work of her predecessors, and return the skin to its unmodified state, or leave it stripy, as it presumably was in life? In 2020 she told me that they had yet to remove the stripes, but the debate was still open. Should they remove something that they know is unoriginal or leave it as it is, since the act of painting it has historical importance too, and does make the specimen appear more realistic?

When Bethany subsequently came across a film of Richard Dawkins using the specimen when he gave the 1991 Royal Institution Christmas Lectures, she spotted that the stripes had not yet been painted on. This means that the undocumented modifications had only been made in the last couple of decades. So what do you think? Should Oxford de-stripe their thylacine?

*

What's the real issue with a wonky echidna, a bipedal wombat or a ballsy thylacine? Everybody loves crap taxidermy (don't take my word for it; there are websites, books and hashtags all dedicated to attaching amusing captions to badly stuffed animals), and so they should. Dead animals can be funny. But it's generally amusing because the taxidermist in question wasn't very good, rather than because they didn't know what the animal they were working on actually looked like. That's when it becomes a problem.

Because Australian animals are so poorly represented in popular culture as a whole, many people are unfamiliar with what they truly look like. As such, museums play a particularly important role in how people encounter these species. This isn't the case with much of the rest of the world's fauna. If people encountered an inaccurate taxidermy monkey, big cat, fox or squirrel, for example, it wouldn't matter so much – they are so prevalent in popular culture, including wildlife films, books and zoos, that the public already has a solid appreciation of what they truly look like in life. That's one reason why people recognise crap taxidermy when they see it: they can instantly spot the mistakes.

By contrast, few Australian animals are given airtime, so when museums display inaccurate taxidermy of species that people are unfamiliar with when alive, they are teaching people *the wrong thing*: unlike comically crap taxidermy, people don't realise that inaccurate Australian taxidermy is incorrect. The visitor then comes away with an entirely wrong image of that species in their mind's eye. It is my contention that – depending on which museum they most recently visited – many people think that platypuses are fat, flat, or sausage-shaped, for example. And that's a shame.

For many, museum galleries are the first place that they have ever looked in real detail at some Australian species. On display, the specimens are acting as ambassadors for their species. That's what specimen means – an example of a wider group.

Most often, display specimens are used as an example to represent every member of their species that has ever lived. It's a big responsibility.

But what do we do when a specimen doesn't look like the species it represents anymore? For extinct species, like faded thylacines, the options are very limited, as we can't get any new ones. But I feel less forgiving about specimens of some other species that have lost their pigments. For example, some museums display Tasmanian devils – jet black in life – that are now a coffee-tinted ash-grey. Or brush-tailed possums, which should have dark grey or even black fur, that are reduced to a ruddy off-white.

Museums today – in a way that they were not in the past – are rightly concerned that it would be unethical to kill a devil or possum to create new specimens to replace faded ones. But collecting skins for taxidermy displays need not have any impact on species conservation. There are plenty of devils in zoos that could become display specimens when they die, and sadly large numbers die as roadkill each year (devils are particular victims to this, as they come to the roads to scavenge the carcasses of all the other animals that have been hit by cars), and institutions there collect these fresh bodies for research. And possums are an invasive pest in New Zealand, constantly being culled to protect the environment. There would be paperwork involved, and costs, but there are ethical routes open for acquiring new specimens.

Tens of millions of people visit natural history museums each year, and I think it's fair to say that most of them don't have a good sense of what Tasmanian devils (again largely thanks to Warner Bros.' long-running misinformation campaign, *Looney Tunes*) and possums look like. Faded specimens are doing their species a disservice. These ambassadors are not successfully representing living devils or possums, and that's a shame. They are not crap taxidermy, they just don't look like the animals they are supposed to, and the problem is that most people won't notice. Colour is only one aspect of their appearance, and there are plenty of other aspects of their biology that visitors will garner from seeing these specimens, but it does trouble me that they will go away with an image in their head that is wrong.

One could also ask if, on balance, the specimens should simply be removed from display, and the gaps filled with different species from the stores that aren't so faded. But that would mean no devils and no possums. Is it worse to teach people the wrong thing about a species, or nothing at all? I don't know the answer.

*

In the minds of evolutionary biologists today, Charles Darwin is obviously king. However, as much as his science is revered as fundamentally magnificent and world changing, I think it's fair to say that many scientists don't find personal connections with him. It's not that he was unpleasant or unscrupulous, it's just that he never really comes off as someone who you'd want to hang out with. A bit of a wet fish.

On the other hand, Alfred Russel Wallace is an absolute legend. Wallace is the much-overlooked co-discoverer of evolution by natural selection (along with Darwin). Self-made and brilliant. Exciting and fun. In the public eye, Darwin gets all the credit, but most biologists – I think – prefer Wallace.

Wallace is undoubtedly one of the greatest naturalists who ever lived. He travelled the world, added mountains of invaluable specimens to museums worldwide, founded entire scientific disciplines based on his interpretations of what he saw, profoundly altered our understanding of the geography of Southeast Asia and Australasia, and wrote the most beautiful prose. To many he is a hero.

I find it hard to criticise Wallace, not only because of his achievements and contributions, but also because he did credit some of the local collectors he hired for their significant roles in his work. One of the most celebrated natural history books of all time is Wallace's *The Malay Archipelago*, a travelogue published in 1869 following his epic eight-year expedition to the region. There, Wallace employed people to join the voyage, to whom he grew very close. A Malay teenager named Ali was perhaps his most trusted expedition member and closest companion, and another named Baderoon also provided instrumental contributions to his collections.[9] Wallace respected them, their insights, their wishes and their Islamic faith, and wrote openly about their part in his accomplishments.

Nonetheless, at times Wallace's writing exemplifies the deep-set colonial view of naturalists and explorers of the time. They felt that when they travelled and discovered, they were seeing things that no one had seen before. The extent to which they discounted the experiences, understanding and knowledge of the Indigenous people living alongside their study animals clearly displays the level of their superiority complex.[10]

In the University Museum of Zoology in Cambridge, we have some of the 8,050 birds Wallace's team amassed on this famous expedition (in total they gathered over 125,000 animal specimens). Each of their labels list Wallace as the collector, but in truth Ali collected many or most of the birds on the voyage and Baderoon was also a key bird-shooter.[11] Among Cambridge's 'Wallace' specimens is a king bird-of-paradise, which is very probably the one that Wallace emotionally describes below, in *The Malay Archipelago*. He proudly states that Baderoon was the collector, but the passage that follows not only illustrates his flair for writing, but also the idea that the Indigenous people who were acting as his guides simply did not count. He and his team had travelled to the Aru Islands between Australia and New Guinea, on an obsessive quest for birds-of-paradise:

I had obtained a specimen of the King Bird of Paradise (*Paradisea regia*) . . . I knew how few Europeans had ever beheld the perfect little organism I now gazed upon, and how very imperfectly it was still known in Europe. The emotions excited in the minds of a naturalist, who has long desired to see the actual thing which he has hitherto known only by description, drawing, or badly-preserved external covering — especially when that thing is of surpassing rarity and beauty, require the poetic faculty fully to express them. The remote island in which I found myself situated, in an almost unvisited sea, far from the tracks of merchant fleets and navies; the wild luxuriant tropical forest, which stretched far away on every side; the rude uncultured savages who gathered round me, — all had their influence in determining the emotions with which I gazed upon this 'thing of beauty'. I thought of the long ages of the past, during which the successive generations of this little creature had run their course — year by year being born, and living and dying amid these dark and gloomy woods, with no intelligent eye to gaze upon their loveliness; to all appearance such a wanton waste of beauty. Such ideas excite a feeling of melancholy. It seems sad, that on the one hand such exquisite creatures should live out their lives and exhibit their charms only in these wild inhospitable regions, doomed for ages yet to come to hopeless barbarism; while on the other hand, should civilized man ever reach these distant lands, and bring moral, intellectual, and physical light into the recesses of these virgin forests, we may be sure that he will so disturb the nicely-balanced relations of organic and inorganic nature as to cause the disappearance, and finally the extinction, of these very beings whose wonderful structure and beauty he alone is fitted to appreciate and enjoy.[12]

'Beauty he alone is fitted to appreciate and enjoy'. Wallace isn't merely saying that the people of the Aru Islands didn't realise the scientific importance these birds would play in evidencing a major strand of evolutionary thinking, but that they could not possibly recognise their aesthetic beauty, nor had the 'intelligent eye[s] to gaze upon their loveliness'. In the artful, talented, engaging sweep of a pen, Wallace has dismissed the perspectives of the people living alongside these species.

This is a different manifestation of colonial attitudes to Indigenous knowledge to the kind on display when Aboriginal Australians were asked whether they knew if platypuses and echidnas lay eggs, but their answers were disbelieved. Here, Wallace is excluding the possibility that the people of Aru value or even have emotional responses to local animals.

Such mindsets cost scientific endeavour – and scientific museum collections – dearly. There are instances of European naturalists engaging Aboriginal people to gather specimens and provide information as a region was being settled. However, in *Australia's First Naturalists*, Penny Olsen and Lynnette Russell explain how the opportunity to learn from Indigenous Australians and incorporate their knowledge into Western science was often squandered in the early years. And as time went on, it became less feasible, up to a point where species had disappeared before they had even been zoologically recorded:

> As cities and major country towns in settled grazing land were established in the south-east and south-west, each new centre afforded opportunities for natural history collectors to find species that were new to them, and the possibility of trade with the Aboriginal occupants. However, around Sydney and other early coastal settlements, the Aboriginal population had become so reduced or separated from country that such assistance could be hard to find.[13]

The European colonisation of Australia came about through the dispossession of its Indigenous population. In part, the newcomers justified it by depicting Australia as a land of inferior inhabitants, be they human or animal. The natural history that was exported back to European scientists was of extraordinary interest, written about prolifically and in ways that fundamentally changed their understanding of the natural world. It is ironic that in neglecting the value of Indigenous knowledge, they made the task of scientific advancement more difficult.

12

Extinction

Throughout this book, I have argued that Australian animals have unfairly been given a primitive reputation. But so what? Why should we care that platypuses, wombats, koalas, bandicoots and kangaroos are regularly slighted by society at large? Am I just being a killjoy? Platypuses *are* weird – should I just get over it?

I don't think so. The argument is quite simple – it's about respect. It's about value. Those 'weird' and 'primitive' tags have consequences for how we value the animals – and the country – they are attached to. It is harder to make the political arguments needed to conserve Australia's fauna because it has been devalued by negative stereotypes – and in turn, the country's human story has been affected by the way the wildlife has been described. What has happened and is happening to the animals and the people is, I believe, entwined with how the animals are represented to the wider world.

Of all the continents, Australia has the best animals in the world. Yes, I know. My opinion is obviously subjective. It is, however, an objective fact that Australia has the world's *worst* record for recent mammal extinctions. Aside from the famously extinct thylacine, Australia has lost a whole host of wonderful creatures. The lesser stick-nest rat, for example, was the smaller of the two rodents that build massive mansions out of sticks, glued together with their urine. The pig-footed bandicoots were the only marsupials to have evolved anything like hooves. Among the others we have lost are rabbit-like rodents that lived in tree trunks and at least five species of jerboa-like hopping-mouse, plus various other hopping specialists – species of wallaby, bettong and hare-wallaby. Since Britain's invasion in 1788, at least thirty mammal species have become extinct – that's nearly 10 per cent of Australia's entire mammalian fauna. In the same period, four of Africa's endemic land mammals have disappeared, five or six of South America's (mostly from tiny islands), none of North America's and none of Europe's.* It's worth noting that all those continents are far larger than Australia. To put it another way, of all the recent mammal extinctions across the globe, 37 per cent happened in Australia.

Since the first of those extinctions, probably in the 1840s, Australia has completely lost one to two species every decade, and that rate appears to be holding true in the twenty-first century so far.[1] And of those species that do survive, many have been reduced to a minute fraction of their pre-European range. Prior to the 2019/20 Australian bushfires, 124 land mammal species were considered to be threatened with extinction in Australia, or near threatened.[2] Those fires were unprecedented in recorded history, and their impacts are yet to be fully assessed, but it is assumed that they will have massively increased the extinction risk of many others. One initial study, for example, found that koala populations in northern New South Wales decreased by 71 per cent as a result of the fires.[3]

What has caused the nationwide environmental catastrophe of the last 200+ years? At the top of the list are introduced carnivores that the Europeans brought with them. Cats were imported both as pets and for rodent control. Foxes,

---

* These numbers reflect the terrestrial species listed as extinct by the International Union for Conservation of Nature, which considers a species extinct once sixty years have passed since they were last seen.

outrageously, were imported simply to be hunted. On top of that, the continent-scale habitat destruction, primarily for agriculture, has been jaw-dropping. It is truly extraordinary how quickly and how extensively Europeans cleared the land for industry and farming. This continues today as weak conservation laws favour industrialists and make it extremely difficult for legal challenges to enforce any protections that might benefit threatened species. Land clearing in Queensland alone – the state with the highest rate of loss of native vegetation – was estimated to kill 100 million native mammals, birds and reptiles each year.[4] Watercourses are diverted for irrigation as well (over half the waterways that feed the largest catchment in the country – the Murray-Darling basin – have disappeared since colonisation),[5] stripping precious water from ecosystems. Alongside this came the introduction of non-native pigs and herbivores (sheep, cattle, goats, camels, donkeys, horses, deer, buffalo, rabbits and hares) for food, sport hunting and transport, all of which have eaten, trampled, buried and pooed on native vegetation and soils to such a degree that few native animals can prosper alongside them. If plants manage to avoid the livestock themselves, these newcomers compress the soil so water runs off it more quickly, changing which plants can live there anyway. Plus, this modified land then holds less water, so droughts hit harder and are more difficult to break.

I've already mentioned several times in this book the impact cats and foxes have had on native wildlife since they were released by Europeans. It is a staggering indictment that they have caused or contributed to most of the thirty or more recent mammal extinctions in Australia. On top of that, they have driven many more species down to minute fractions of their former ranges. A study in 2018 assessed Australia's remaining land mammals (excluding bats), and found that sixty-three were either 'extremely' or 'highly' susceptible to predation by cats and foxes.[6]

The sheer scale of the devastation caused by cats in particular can be overwhelming to consider. The numbers are impossible to comprehend in any meaningful way, but of all the statistics I have come across in my career, those that relate to what cats are doing to Australia's wildlife are the ones that shock me the most. Prepare yourself . . .

In an average year, there are nearly three million feral cats living in Australia. Between them they kill 512 million native mammals a year.[7]

Five hundred and twelve million.

512,000,000.

Over half a billion.

Every year. And that's just the native mammals. Let's convert that to a daily total: every morning, Australia wakes up having lost 1.4 million native mammals that were alive the day before, just thanks to feral cats.

If we add to that the native birds and reptiles killed by feral cats, the annual death toll reaches nearly 1.4 billion.[8]

Each individual feral cat living in the bush (the numbers are lower for feral cats living around human settlements) kills on average 221 native mammals each year. One living cat – 221 dead mammals. It can't go on. Controlling feral cat numbers is a central part of conservation planning in Australia, and in 2015 the government announced its aim to cull two million of them by 2020.[9] By the middle of 2018, they reported that 844,000 had been killed – a big number, but falling shy of their target.[10] Experts in the field argue that the two million goal was a media-grabbing political move intended to draw attention away from the scale of the more polarising subject of Australia's habitat destruction. They say it was not based on any scientific assessment of the impact of the cull, for example how readily cat populations can replace their numbers, or how individual culls were spread around the country in relation to where threatened mammals live.[11]

And all that's just about the feral cats. There are an esti-mated 3.8 million pet cats in Australia, which destroy a further

230 million native birds, mammals and reptiles each year (of which 67 million are mammals).[12] I think that number should be attached to the collar of every pet cat in the country, in the hope that cat owners will appreciate the importance of neutering their cats and keeping them indoors, particularly overnight, or confined to garden 'cat runs'. All those people that say, 'Oh, my little angel doesn't kill wildlife' simply cannot be correct (studies have shown that on average cats only bring home 15 per cent of the animals they kill).[13]

Let's not forget, either, arguably the worst-informed deliberate animal introduction in human history: the cane toad. These toads were taken to tropical Australia on the gamble that they might control the economically damaging cane beetles, which are pests on sugar cane crops.

It was a disaster for two predictable reasons. First, cane toads are capable of eating nearly everything that's smaller than them. As they can reach 30 centimetres (1 foot) in length and can well exceed a kilogram (2 pounds) in weight, a lot of animals are smaller than them. And second, they poison nearly everything that's bigger than them, thanks to defensive toxins they have in their skin to ward off predators that try to eat them. Among the cane toads' victims are monitor lizards (which are poisoned by the toads) and predatory ants (which are eaten by the toads). Both are significant predators of cane beetles themselves: the toads killed the animals that were keeping the beetle numbers in check.[14] So, in the end, the cane toads did not improve the sugar cane crop yield (and may have made it worse), but they have had an extraordinary environmental impact.

Initially, 101 toads were imported from Hawaii, bred, and their offspring released in a spot near Cairns, in the northeast, in 1935. A female is capable of laying 30,000 eggs per clutch, twice a year, and they now occupy over 1.5 million square kilometres (580,000 square miles) of Australia and continue to spread unabated. They are so prevalent in northern Australia that on one field trip I found nineteen toads in a single bucket-sized pitfall trap after one night.

Disgust is not an emotion one commonly finds among ecologists. It's fair to say that it takes a lot to turn our stomachs during fieldwork. I remember thinking it odd, for example, that my reaction wasn't more visceral when I had to kneel directly in the open stomach of a disembowelled cow that had been rotting for a week. It was the only spot I could safely position myself to handle a dingo we had trapped in order to fit it with tracking tags. However, unquestionably the most repulsive experience of my life was preparing bait for a study into whether northern quolls could be trained not to eat cane toads (these predatory marsupials are one of the species worst affected by toads). My friend Georgia Ward-Fear had demonstrated that monitor lizards could be encouraged to learn not to eat toads,[15] and researcher Naomi Indigo was attempting something similar with quolls. It involved trying to get wild quolls to associate the taste of toad with feeling ill, so that when they encounter a killer toad while hunting, they don't eat it and die.

To do this, I was helping Naomi convert a cubic metre (35 cubic feet) of dead toads into non-lethal sausages. This was the day that my disgust-threshold met its match. Toad poison is produced by their skin, and the very largest glands are by the back of their heads. To make sausages that might make quolls feel queasy, but not die, we avoided the most toxic parts, blending the meat from their legs with a very small amount of leg skin. The process involved bisecting the toads at the waist, skinning their legs (which is surprisingly easy – it mostly comes off like a pair of leggings, although getting it cleanly over the feet requires a good grip on the greasy, elastic skin and a deft, hard, snapping yank) and mincing the appropriate bits to a specific ratio. Handling dead animals and their constituent parts is all part and parcel of my day job, but as that day wore on, the tropical heat's influence on the volume of toad meat generated a stench in our makeshift field lab that I will not forget (thankfully this smell also made the sausages more detectable by the quolls). It was worth it. Naomi's research did indeed show that quolls that ate her sausages were unlikely to come back for a second helping, raising hope that they will remember the taste as clearly as I remember the smell, and leave whole toads alone in the future.[16]

As well as the menagerie of deliberate additions to the Australian fauna and the devastating land clearances, other major causes of biodiversity loss include the rats and mice that were inadvertently introduced, climate change (the world's first mammal to definitively be wiped out as a result of human-induced climate change was an Australian rodent, the Bramble Cays melomys, in 2015, which succumbed to rising sea levels on its low-lying island home) and the modified fire regimes that result from it, and the enforced changes to Aboriginal Australians' management of the land. The last two in that list are linked. Australia is getting drier, and the impacts of bush fires are intensifying. At the same time the decline in regular, low impact fires that only affect small patches at a time – exemplified by traditional mosaic burning practices – leaves enormous tracts of land vulnerable to massive fires running uncontrollably. It's a sad fact that Australia is a bad place to be for Australian mammals, which is problematic because for the vast majority of species that is the only place that they live.

*

Whether species disappear completely, or from a part or most of their range, the world turns differently as a result of these extinctions. Without their native mammal faunas, entire ecosystems are collapsing.

Without digging animals like bandicoots, bilbies and bettongs, the underlying chemistry of the landscapes can change for several reasons. First, the process of turning soil over, as any gardener knows, has an impact on what will grow in it. Second, when these animals dig for the insects or fungi they eat, they leave little divots in the soil. This reduces the monotonous flatness of the landscape in an apparently small, but critically important way. When the wind blows, leaf litter collects in these little pits. So, bilbies and bettongs create countless miniature compost heaps wherever they go. As the leaves rot, they alter the amount of carbon and nitrogen available in the soil for growing plants, particularly in the drier parts of Australia.[17]

There are other ecological changes, too, which we are just beginning to understand. In Australia, ants and birds were considered the main predators of seeds – and therefore major players in plants' normal lifecycles – and the impact of mammals was considered far less important. We now know, however, that the period when scientists decided what 'normal' looked like for Australia was *after* mammals had already largely disappeared from the landscape as a result of European invasion, and therefore was not normal at all.[18]

In a few tiny pockets of Australia, dedicated conservationists are beginning to reintroduce animals that have been lost. For example, at the Arid Recovery wildlife reserve in central South Australia (where I met my first bilby), ecologists have fenced out cats and foxes and brought some of the missing animals back. In the absence of introduced predators, the small population of bilbies they reintroduced there at the turn of the millennium has grown to over a thousand. And in those twenty years, the twenty-nine-strong founding population of burrowing bettongs exploded into the thousands. Scientists now visit this reserve halfway between Adelaide and Alice Springs to understand what the ecosystem might have looked like before the recent centuries of change.

They are finding that the little bilby compost pits are encouraging nutrient cycling, and that the bettongs are eating the seeds of the choking woody shrubs that have been encroaching across Australia's arid zone, enabling grasses and smaller, softer plants to re-establish. With different plants, the whole landscape changes. As more areas of Australia are rewilded in this way, conservationists are bringing about a tiny ray of hope for the future of the ecosystems.

This work is inspiring. However, it is hard not to sense the profound loss that Australia has suffered. I am in my thirties, and there are Australian mammals that no longer exist today but were alive when I was born. The scale of loss is increasing, too – not just because more species are going extinct now, but also because we are repeatedly discovering that more species were present at the time of European invasion than previously thought. We've lost more than we realised: animals that we assumed were single species that went extinct in the recent past were actually multiple species. As I mentioned previously, taxonomists studying museum specimens of pig-footed bandicoots collected long ago announced in 2019 that what had always been assumed to be a single species by Western scientists was actually two species – both gone.[19]

One of those scientists – Kenny Travouillon, at the Western Australian Museum – led another study published the year before, which found that the species we knew as the western barred bandicoot was not one species, but five.[20] Western barred bandicoots once lived over much of southern Australia, but by the 1930s had been driven to extinction on the mainland, primarily due to introduced cats and foxes. They survived only on islands in Shark Bay. In recent decades, some of these survivors were transported to establish new mainland populations in fenced sanctuaries where feral predators have been eliminated – Arid Recovery in South Australia and the Australian Wildlife Conservancy's Mount Gibson reserve in Western Australia. As a result of this new study, however, we now know that those bandicoots holding out in Shark Bay were not the same species as most of those that had disappeared from elsewhere in Australia. While redefining what constitutes a western barred bandicoot, Kenny and his colleagues described four new, closely related species, all of which are extinct. This means that the bandicoots taken to Arid Recovery and Mount Gibson were introductions, not reintroductions.[21]

Aside from mourning the loss of four further species, conservationists are not overly worried about this translocation, as despite apparently being slightly different from their relatives that were originally found in the arid centre, the Shark Bay bandicoots are performing the same vital habitat-engineering roles that their extinct cousins once did. As Katherine Tuft, who manages Arid Recovery (and who took me on some of my first ecological field trips in Australia, and many more in the decade since) says, 'Surely it is better to have the "wrong" bandicoot than no bandicoot at all.'[22]

*

As Arid Recovery and other rewilding projects teach us, even when we look at the apparently 'wildest' parts of the Australian outback, what we see there does not reflect pre-colonisation ecosystems. William Caldwell, when he was collecting embryos in the 1880s for his research into how marsupials develop, tells stories of catching thousands of kangaroos at once – but he recognises that he may be among the last to do so:

> The kangaroos have decreased in number . . . and the place I am in now is, I believe, almost the only one where it is still possible to get a thousand kangaroos into a "yard" in one day. "Yarding" has been generally superseded by shooting. A camp of kangaroo-shooters will travel about on a run for months, being paid so much a scalp. It is very slow work collecting embryos with these shooting-parties, and, besides this, the embryos are too delicate to be carried on horseback. Accordingly, I have tried hard to get to a

yarding-drive where I could put up a table and do all the preserving in one place. On Tuesday, Wednesday, and Thursday next week the whole district is going to muster to drive kangaroos into a pit, and we hope to get five thousand.[23]

Such a concentration is not the Australia I know 140 years later. I can't remember seeing more than fifty kangaroos at once.

Aboriginal hunting may have dramatically altered the Australian ecosystems with the disappearance of the megafauna, but when it comes to thinking about species harmed by hunting since European colonisation, it is the thylacine that most often springs to mind. Obviously, the outcome was absolute for the thylacine – it was shot, snared and trapped out of existence – but we should also think about how platypus populations have been modified by hunting.

Perhaps surprisingly, in the late nineteenth and early twentieth century platypus fur was more valuable than that of any other Australian animal.[24] Their skins were desirable as they are glossy and dense (the cold-adapted Tasmanian skins are the most luxurious) and relatively uniform in colour and length, making them easier to stitch together. They are stiff, however, so they were chiefly used in making rugs, and one rug would require the skins of fifty to eighty animals.

Despite the generous private and government-sponsored bounties, hunting thylacines was not profitable enough for there to be any dedicated professional thylacine hunters.[25] By contrast, a living could be made from killing platypuses, and individual hunters shot thousands of them over the course of their careers.[26] In order to kill the platypuses without damaging their skins, heavy calibre shot was fired into the water below the animals, which would stun them, and then dogs were sent in to retrieve them (which were occasionally stung in the process).[27]

Today, we think of platypuses as solitary – you'd be very

lucky to see more than one at once – but Harry Burrell gives a remarkable account that may suggest things used to be different:

> A migration of platypus was observed by the late Mr. William Hill in 1859, when he was manager of the Pallamallawa cattle station (now a township) on the Gwydir River. About fifty aborigines, under 'King' Binamoore, were camped on the river bank not far from the homestead. One evening Mr. and Mrs. Hill strolled down to the camp for a yarn with Binamoore, who was an old friend, but the chat was interrupted by a sound from up river, similar to that made by a mob of cattle fording a stream. . . . Presently the noise was heard again, this time closer at hand, in fact so close that Binamoore and his tribe, and the tribal dogs, took fright, and bolted off to the next station, deserting their gunyahs [huts]. Mr. and Mrs. Hill then clearly observed a mob of platypus all swimming together at top speed with the current, and estimated that there were at least a hundred of them.[28]

If true, the idea of a group of platypuses (apparently a 'puddle' is the collective noun) so large that the sound of it could be mistaken for a herd of cows blows my mind. In any case, accounts of the scale of the platypus fur trade suggests that it occurred at such a rate that the populations we see today must be significantly smaller than they were during the time of the trade. These population changes were recognised by the authorities, as platypuses became legally protected in all their range states by 1912.*

Human hunting no longer presents an existential threat to platypuses, and writing in the 1980s, platypus biologist Tom

---

* Victoria: 1892; New South Wales: 1901; Queensland: 1906; Tasmania: 1907; South Australia: 1912 (Bino et al., 2019).

Grant stated that although numbers have been reduced by human activity, particularly around settlements, 'we can say that at present *O. anatinus* is not a rare or endangered species but is in fact a common inhabitant of most of the streams and rivers of eastern Australia, from Cooktown in the north to Tasmania in the south'.[29]

Nonetheless, their numbers do not seem to reflect those of the recent past, and several leading researchers today think we ought to be far more concerned about the plight of the platypus. In 2016, the International Union for Conservation of Nature – the organisation responsible for assessing the scale of the threats faced by species worldwide – upgraded the platypus from its lowest category, 'Least Concern', to 'Near Threatened'.[30]

The variety of threats platypuses face over their range is vast (and not particularly unusual for an Australian species). In a major 2019 review of platypus ecology, Gilad Bino and his colleagues listed the dangers they face:

> Distribution of the platypus coincides with major threatening processes, including highly regulated and disrupted rivers, extensive riparian and lotic [riverbanks and fast-moving freshwater] habitat degradation by agriculture and urbanization, and fragmentation by dams and other in-stream structures. By-catch mortality in fishing gear, diseases, and predation by invasive foxes and feral dogs* also impact platypus populations.[31]

On top of all that, climate change is also a factor in their ability to survive, as temperatures are rising beyond those that platypuses can tolerate.[32] Despite this litany of problems, platypuses are not currently on the threatened species lists of

---

* One study found that 40 per cent of platypus deaths in Tasmania were caused by dogs. (Connolly et al., 1998)

most of the states in which they are found, nor are they considered at risk by the federal government. Experts want this to change and have recommended that they are considered vulnerable to extinction both nationally and in Victoria and New South Wales. Victoria heeded this advice in 2021 and joined South Australia by including platypuses on the state's threatened species list.[33] This follows a 2020 study that found that over the past three decades alone, the area occupied by platypuses had shrunk by over a fifth.[34] Australia would be immeasurably poorer if it were to add the platypus to its long list of lost mammals.

*

All in all, the overriding cause of this conservation emergency is that the Australian government has consistently failed to sufficiently protect its native wildlife. Australia has no legislation on its statute books that obliges the government to actively protect its threatened species, and so it doesn't. Australia has powerful industrial lobbies – for mining and mineral extraction, and for farming. These special interests have far more political power than possums and platypuses.

In 2020, a ten-year review of Australia's Environment Protection and Biodiversity Conservation Act – the major piece of national legislation which supposedly safeguards species and ecosystems – found it to be ineffective, and that very little had been done to enforce it over the twenty years since it has been in place.[35] Essentially, it creates laborious and inconsistent processes for how to assess whether species or habitats are threatened, particularly by major industrial developments like coal, gas and mineral extraction. If species *are* found to be at risk, plans have only rarely been developed for how to help them recover, and the Act makes no requirement to do so.

The review found that environmental laws in Australia were

rarely policed, and when they were, the penalties for breaking them were minor. The combined fines issued to developers who failed to deliver the environmental safeguards they had committed to were lower over the course of a decade than the parking fines individual local authorities collected in a single year.

Neither environmentalists nor developers had been satisfied with the Act – one group saying it failed to protect the environment, and the other saying it created too much red tape and delayed projects. Over the course of two years, journalists at *Guardian Australia* investigated its systemic failings and were thoroughly damning of

> widespread problems, including poor monitoring of endangered species, major delays in the listing of threatened species and ecosystems, failure to develop, update and implement recovery plans for species and habitats threatened with extinction, failure to list key threats to species, failure to protect important habitat, and threatened species funding being used for projects that do not benefit threatened species. The laws also do not address the effects of climate change.[36]

I should be clear that this is not an indictment of the amazing work that conservationists are undertaking in Australia – far from it. In fact, their efforts and successes are even more impressive as they are achieved against a backdrop of insufficient federal support.

The independent review recommended that the government develop a set of legally enforceable standards that make clear the rules for environmental protection, and an independent environmental regulator to monitor and enforce the laws.

Initially, the government appeared to commit to adopting the standards but refused the idea of a regulatory body.[37] However, it then introduced legislation to give control over planning approvals for potentially environmentally destructive developments to individual states and territories, in order to cut approval times and promote economic growth – particularly in response to the 2020 coronavirus pandemic – but without reference to new environmental standards. Conservation groups were left feeling that despite the scathing review, the changes that the government then proposed risked making the extinction crisis even worse. Further, in 2022, the State of the Environment Report concluded that in all categories but one, Australia's environments were rated as deteriorating, and most were 'poor and deteriorating'.[38] Successive governments have failed in their responsibilities.

In my view, all this is tied in to the way Australian animals are represented to the outside world, and in Australia itself. I doubt we will ever be able demonstrate a definitive link between cause and effect – for instance, we won't be able to say that a given population of animals disappeared thanks to a specific slur on marsupials. However, it is reasonable to assume that species that are valued less do not enjoy the same prioritisation when it comes to environmental protection. Consider a realistic example: on the one hand, there might be developers who are lobbying to clear miles of native vegetation to increase profits. On the other, there are the animals whose habitat would be destroyed in the process. When a legislator is deciding whether to give the go-ahead to allow the bulldozers in (or even whether planning permission is required in the first place), their attitudes towards the wildlife in question will have an impact on how they balance one side against the other. The same is true for the landowners who had wanted to raze the land in the first place. The more people who value a species, the safer it becomes. And so, as long as people continue to incorrectly refer to Australian wildlife as merely a weird bunch of primitive curiosities, inevitably doomed to be outcompeted by a superior evolutionary force from the north, then conservation is unlikely to be prioritised. If we add to that the incorrect, but popular, idea that these animals are less intelligent, then the ill-informed might suggest that they are 'too stupid to survive' when they are left to compete with livestock and their feral counterparts, as well as rabbits, invasive rats and mice, foxes and cats. Australian mammals are devalued by the way we talk about and represent them in everyday language, museums, popular culture and scientific research, and this is having a catastrophic impact that inevitably contributes to the extinction crisis.

13

Terra Nullius and Colonialism

The way Australian wildlife is portrayed hasn't just had an impact on its own perceived value. It goes further than that. The colonial idea that Australia was fair game for European imperial occupation and settlement was based on the concept that it was *terra nullius*, 'nobody's land', as the bigoted European invaders decided that the Indigenous civilisations they encountered were too savage, uncivilised and primitive to have true ownership.

This is why the platypus and other Australian mammals matter. And why the way we talk about them matters. I believe that the perceived status of the people and the animals in Australia were fundamentally intertwined in the minds and the words of the colonisers. It served their political narrative to dismiss both people and animals as inferior, because it augmented the arguments to justify invasion. One 1840s settler set out their case by giving a human persona to their animal enemy, 'If the kangaroo are allowed to live and multiply, our sheep will starve. We can't live if they don't. Ergo, it is our life and welfare against Marsupial Bill's, and he, being of the inferior race, must go under'.[1]

By tying animals and Aboriginal people together in an alleged collective inferiority, it became easier to paint Australia as a primitive, degenerate backwater. Through their denigrative written descriptions, the imperial establishment created a hierarchy in which Europe was made to look superior to Australia in every respect – the people, the animals and the climate. If that's all there was to Australia, then how could anyone object to Britain taking possession of it and overseeing its 'improvement'? There's a sense that some even felt that the colonisation of Australia wasn't an invasion at all, but an act of benevolence.

It's hard to overstate the horrific injustices suffered by Indigenous Australians at the hands of colonial invaders, who robbed them of their land and much more. The 'natives' were considered an inferior race in need of Western civilisation – to be implemented through force. Furthermore, Aboriginal and Torres Strait Islander communities continue to be thoroughly institutionally marginalised in modern society. It isn't difficult to demonstrate that the structure of the Australian constitution continues to be systemically racist today.[2]

The parallels between human and animal treatment in Australia are disturbingly close. To a great extent, both the cultures and the ecosystems of Australia have been overwritten; people and animals have been dispossessed of their land. For instance, just like what has happened to the people of Australia, European acclimatisation societies methodically sought to replace the fauna and flora of the land with species they were accustomed to seeing back home. This isn't about the large-scale clearing of habitats that took place for farming livestock. Acclimatisation societies were local organisations across new colonial settlements dedicated to bringing a sense of comforting suburban England to the colonies by introducing familiar British species. They were also driven by the notion that their new home was faunally impoverished, and that the European species they let loose would improve the landscape, again reflecting attitudes towards what was happening with Australia's new and existing human inhabitants.

Instead of enhancing ecosystems, however, these introduced species have caused untold damage to them. Rabbits are now one of the most financially costly alien species in Australia, for example. Walk through the suburban streets in many parts of Australia and you are likely to spot European starlings, house sparrows, rock doves, Eurasian blackbirds and goldfinches, to name a few of the birds the acclimatisation societies brought over.* Visit the parks and you may spot a mallard. Many animals and plants were introduced purely to make Europeans feel more at ease, while native species were displaced or actively eliminated. This mirrors the treatment of Indigenous Australians.

Movements of wild mammals in the opposite direction have been conspicuously rare. Over time, a few colonies of Bennett's wallabies – native to Tasmania – have established in the UK, including in England's Peak District (where they are now probably extinct) and on an island in Scotland's Loch Lomond, for example.† And when a pair of brush-tailed rock-wallabies escaped from a tent in Hawaii in 1916, a population became established there. This has even been discussed as a possible source for reintroducing animals back to Australia where the species is now in decline.[3] While people have moved several species around within Australia (causing damage to the places they do not belong), or taken them from Australia to New Guinea, New Zealand and other Pacific islands, it seems very few have been introduced to countries any further away, and even fewer of those were released deliberately.

Rather than introductions of Australian animals into the wilds of Europe and America, instead we might consider the role of zoos in the translocation of both animals and people

---

* It's worth saying that not every bird species that is spotted in both Australia and Europe is introduced. Aside from many migratory water birds, the same species of coot, great-crested grebe, barn owl, peregrine falcon and osprey, for example, are native to both regions, as well as the Americas, Africa and parts of Asia.

† Mention of Loch Lomond allows us to note that the colonists didn't only import animals into Australia, but also place names. Ben Lomond – the mountain that sits above Loch Lomond – is one of many British names that was reused in Australia: Ben Lomond is also a mountain in Tasmania. Needless to say, these locations already had names when Europeans overwrote them.

in that direction, and the ways in which they were displayed and described there that diminished their statuses. Australian Aboriginal people – at the same time as Australian animals – were transported and degradingly exhibited as spectacles in 'human zoos' across Europe, as just another of the spoils of empire.

Narratives insisting on the equality of humankind were not sufficiently prominent while these injustices were playing out, nor indeed was exploration of Aboriginal Australians' significant accomplishments. In reality, genetic studies in recent years have confirmed what Indigenous Australians have long known; that they are the longest-living culture in the world – descendants of the first real human explorers, with the nerve and technology to have been the first to cross oceans.[4] In his controversial history of Australia, *Dark Emu*, Bruce Pascoe makes the point that archaeological evidence of sophisticated Australian encampments 'have proved to be the oldest in the world, a discovery that suggests Aboriginal people also invented society.'[5]

From the outset, the ways that traditional Indigenous cultures have been framed in colonial accounts of Australia have had profound impact on First Nations peoples' rights in their own country. The people the colonists encountered were described as skilled hunters with in-depth knowledge and understanding of how to survive in Australia's challenging outback. They could read the land, find food and water, anticipate the seasons and navigate over vast distances. Indeed, many European explorers would have died if it weren't for their assistance. On the face of it, these are positive portrayals, but the politically important part is that they lived nomadic lives as hunter-gatherer-fishers. An idea was used against them that nomads don't own land; they just move over it; hunter-gatherers know how to survive on the environment's bounty, but they don't consciously manage it. They are passive beneficiaries, rather than active agents who have deliberately manipulated nature for societal gain. As one *Sydney Herald* editorial from 1838 put it, 'They bestowed no labour upon the land and that – and that only – it is which gives a right of property to it'.[6] This was a fundamental pillar of the *terra nullius* argument, but it is wrong.

The patronising idea that hunter-gatherers passively wandered the land living simple lives misses the complexity of these peoples' cultures and their level of agency in managing country, their diets and traditional practices. These societies were actively managing the land in order to encourage prey and create the surplus food stores that are necessary to feed a large population beyond seasonal harvests. The evidence for the comprehensive management of the entire continent is laid out in extraordinary detail in Bill Gammage's 2011 masterpiece *The Biggest Estate on Earth*. While today much of Australia is considered remote, inhospitable, untamed and wild 'outback', as Gammage put it, before Europeans arrived, 'there was no wilderness'.[*]

He demonstrates that Indigenous Australians used fire to create different habitats for different reasons, with scalpel-like precision. Fire management was planned in exquisite detail, taking place at different times of year depending on the desired outcome, with rotation cycles that could operate in timescales over generations: some areas were burnt annually, some every few years, and sometimes decades would pass between burns – but these were active, timed decisions: their use of the land was not passive.

In their accounts of the country, settlers variously described habitat features such as lawns, tree rows, clumps, coppices, plains, belts and clearings, often arranged in mosaics that were surely deliberately created. Most notably, the earliest newcomers to many regions consistently wrote about finding 'parklands'

---

[*] On this topic, I am reminded of an article that warns Western ecologists about referring to their field sites as 'remote' as a generalisation, as they may only be 'remote' from an urban, Eurocentric or North American perspective, and they are not remote to the people who live there (Baker, Eichhorn & Griffiths, 2019). Both 'remote' and 'wilderness' in this sense can be colonial terms.

– open grasslands with single trees scattered very infrequently. They thought they resembled gentlemen's parks back home. Not only did they remark on their beauty, but also that verdant pastures like these, with few obstacles to horseback riding, were perfect for grazing livestock. Almost all assumed that these landscapes were natural. The irony is that had the land not been so well managed, 'the invaders might not have come'.[7]

Curiously, as Gammage demonstrates, time after time these grasslands were found on the most fertile soils, with trees often growing in number only in the least fertile places. This is the opposite of what they would have expected, based on what happens in Europe. The reason is that Aboriginal people were managing the fertile land for productivity:

> They first managed the land for plants. They knew which grew where, and which they must tend or transplant. Then they managed for animals. Knowing which plants animals prefer let them burn to associate the sweetest feed, the best shelter, the safest scrub. They established a circuit of such places, activating the next as the last was exhausted or its animals fled. In this way they could predict where animals would be. They travelled to known resources, and made them not merely sustainable, but abundant, convenient and predictable. These are loaded words, the opposite of what Europeans once presumed about hunter-gatherers.[8]

Despite the similarities they repeatedly listed between privately owned, exclusive gentlemen's parks – which were obviously created deliberately – and what they found in Australia, almost no one made the connection that they, too, had been purposefully made: 'Almost all thought no land in Australia private, and parks natural. To think otherwise required them to see Aborigines as gentry, not shiftless wanderers. That seemed preposterous.'[9]

Europeans saw that Aboriginal people were using fire in these habitats to hunt, but not that they had made them that way in the first place. They described the burning, but rarely the level of planning and management required, over generations, to maintain the country. Soon after Aboriginal management ended, the 'parks' became tangled with undergrowth – useless to anyone. Today, it is becoming increasingly common for modern fire management to be informed by traditional practices in order to open up country that has become choked with scrub since dispossession.

Indigenous Australians were also using sophisticated technologies. William Dampier, one of the first Europeans to make detailed observations of Aboriginal life when he landed on the Kimberley coast in 1688 aboard the *Cygnet*, described the weirs that the people there built to act as tidal fish traps, and suggested that this was their principal source of food.[10] These weirs are an ingenious system for catching large numbers of fish with little ongoing effort: a wall is built – out of wood, thatch or stone – across the shallows of a bay. When the tide goes out, fish get trapped behind the wall as the water level drops, eventually draining completely away to leave the fish strewn on the sand behind the wall. Once the wall is built, all you have to do is walk along it twice a day and collect your harvest.

Elsewhere in Australia, Pascoe describes elaborate fish traps with interlocking stones built over rivers and lakes, or dug into artificial lagoons branching off them, reshaped with dams and dykes; and finely crafted nets reaching 270 metres (885 feet) long. Others constructed meshed fences over rivers that funnelled fish down into narrow openings to which nets would be attached whenever food was required. They were extending fishes' natural range, too. Aqueducts were raised and canals cut to introduce eels inland over watersheds, with some constructions requiring decades to complete.[11] At Brewarrina in northern New South Wales the Ngemba people engineered an extraordinary system of ponds and fish traps to manage a massive sustainable harvest (and which had holes small enough to allow the young fishes through, to ensure there would be enough food for other communities upstream). It is plausibly suggested to be the oldest human construction on the planet, with one study dating it to 40,000 years old.[12]

In certain parts of the country, early colonial explorer diaries describe places where Indigenous Australians stayed (sometimes for months at a time) as having well-trodden tracks and being near vast areas where native grains and tubers grew, alongside fruit trees and stores of milled grain and smoked fish to last the leaner months. Indeed, the European explorers who described them stole the food stores they found (and for some reason burned camps they passed through).

The native grains and tubers that some First Nations groups utilised had evolved to thrive in Australian environments. Surely it is worth investigating whether there is commercial value in resurrecting these traditional practices in place of the technologies and crops that were developed for Europe yet have overwritten the Australian systems and may be ill-suited there. Much of Australia has proved challenging to farm in the European way. Gammage suggests that Aboriginal people harvested these plants over more areas of the country than Western-style agriculture is today. Native species are likely to require fewer pesticides, less water and less irrigation, for example.

As Pascoe points out, it's curious that the typical food used to represent Indigenous diets in 'bush tucker' experiences is witchetty grubs, rather than two major groups of plants they used – yams and grains, 'almost as if there is a deliberate attempt by educationalists to emphasise the gross and primitive'.[13]

The racist *terra nullius* argument insisted that ownership required active land-management, but didn't actually pay attention to whether this was taking place. The West found it convenient to ignore evidence of the complexity of how Indigenous civilisations lived off the country, and their profound success. Just as it suited a colonial strategy to downplay the value of their native wildlife, the occasionally romantically portrayed notion of 'primitive' nomadic hunter-gatherers better suited the colonists' narrative to claim it was unowned land. Gammage argues that we could flip this:

> Only in Australia did a mobile people organise a continent with such precision . . . They were active not passive, striving for balance and continuity to make all life abundant, convenient and predictable. They put the mark of humanity firmly on every place . . . This is possession in its most fundamental sense. If *terra nullius* exists anywhere in our country, it was made by Europeans.[14]

A debate has been running in Australia over whether traditional Indigenous land-management should be considered farming. Gammage sets out a series of accounts that, through time, consistently attempt to exclude Aboriginal Australians from definitions of farming. A requirement of farming would be put forward, and every time evidence was found that showed Australians met it, the goalposts would be shifted to present a new requirement. However, in their critique of Pascoe's *Dark Emu*, Peter Sutton and Keryn Walshe make clear that this rather misses the point.[15] The problem is that it implies that farming is superior to being a hunter-gatherer. Several academics have used language like 'proto-farming', 'incipient agriculture' or 'they were on their way to farming'. It's a deficit model that suggests that if only they were given enough time, eventually Australia would 'progress' to develop fully-fledged farming. It's the *terra nullius* argument all over again, but disguised in the form of social Darwinism. Societies do not evolve on a single track from 'primitive' hunter-gatherers to 'advanced' farmers. This is an ego-driven lie the West has told itself. The astonishing complexity of First Nations Australian culture should not be considered inferior. It doesn't matter if we call it farming: their way has worked for tens of thousands of years, and should be valued in and of itself. The invention of a hierarchy of societal progression is just another manifestation of colonial thinking.

A chief difference between European and Australian systems was that the latter allowed people to be mobile. Settling in one area is assumed to be a requirement of farming, as societies need to guard their crops and stores, often using physical force. Instead, Aboriginal Australians used religious sanctions as safeguards against loss to their neighbours. Theft was very rare as a result, enabling them to move away for a period if it were beneficial to do so. This mobility and adaptability made them less susceptible to the whims of drought than European farmers: there was no need to stay put.

But this is what many Europeans found so distasteful, so different and so hard to accept. Gammage explains how Aboriginal management and mobility made food acquisition at least as predictable as farming, and more so in times of drought and flood. They didn't need to work so hard to feed themselves, and this afforded them time for leisure, religion and ceremony. 'These were the preserves and pursuits of gentry. It did not seem right that Aborigines should be like that. Aborigines were "shiftless and improvident", uncivilised. The words meant to degrade hunter-gatherers'.[16] They were viewed with prejudice and scorn as an outcome of the fact that they managed their land so sustainably.

Looking at how Indigenous heritage is treated today in Australia, perhaps it isn't surprising that sufficient respect for the true nature of historic Aboriginal cultures hasn't become firmly established in the national zeitgeist. As I was writing this chapter in May 2020, the news broke that the global mining corporation Rio Tinto had, under a legal permit issued by the Western Australian government, used explosives to blow up two of the most important archaeological sites in the country, in order to expand one of their iron ore mines. With evidence of occupation dating back as far as 46,000 years, the rock shelters at Juukan Gorge in the Pilbara region were among the country's very oldest sites. They were used throughout the last Ice Age – the only site in the state to show continued occupation.[17] The oldest example of bone tools in Australia had been found there, as well as the oldest grinding stones for processing cereals and seeds in Western Australia. The destruction was compared to Islamic State's demolition of Palmyra in 2015.[18] That horrific attack on global heritage quite rightly made significant headlines the world over, but when a mining giant obliterated a site more than 40,000 years older than the pyramids of Giza for the sake of some iron-rich dirt, the news barely made it out of Australia. The idea that exportable dust is more valuable to the Australian government than its earliest human heritage suggests that more work needs to be done to finally be rid of the lie of *terra nullius*.

*

The first formal colonial outposts on the island state of Tasmania were established in 1803, marking the beginning of parallel stories of concerted environmental destruction – and an act of genocide.* Conflict between Indigenous Tasmanians and settlers inevitably broke out as their land was claimed for farming, and kangaroos – a key source of protein for Tasmanian people – were unsustainably harvested by the rocketing numbers of newcomers. It didn't take long for Europeans to significantly outnumber Tasmanian Aboriginal people.

The period from 1824 to 1831 has become known as the Black War. Lieutenant Governor George Arthur followed a declaration of martial law – which allowed the killing of Indigenous people with impunity – with an active bounty in 1830. It rewarded settlers for the capture or killing of

---

* Somehow the imperial propaganda machine largely suppressed the notion that by deliberately setting out to extirpate Tasmania's Indigenous population, the British were committing genocide (Harman, 2018).

Aboriginal people. This was the very same year that the first bounty was established for the killing of the thylacine, initially arranged by the Van Diemen's Land Company – a farming corporation still in existence today.* The congruence could hardly be more poignant. The government added their own thylacine bounty in 1888.

Thylacines were unfairly blamed for killing large numbers of sheep. However, among the true culprits were dogs that had been bred by the settlers for hunting kangaroos, which had been allowed to go feral. In this way, kangaroo hunting by Europeans and their dogs is a root cause of the fate of both the thylacine and Aboriginal Tasmanians.

Also in 1830, Arthur ordered the largest ground offensive in Australian colonial history: an astonishing military operation known as the Black Line, when over 2,000 armed Europeans marched across the state in a coordinated cordon in an attempt to kettle the Indigenous population onto the tip of the Tasman Peninsula. It was predicted that they would push 500 Aboriginal people into the region, but they only managed to catch a tiny handful, although they shot dead many more.

A supposedly more conciliatory strategy started in 1831, under the management of businessman-turned-missionary George Augustus Robinson, when fifty-one Tasmanian Aboriginal people were persuaded to relocate to Gun Carriage Island – now called Vansittart Island – in the Bass Strait. They were told they could live there in safety without fear of attack from the white invaders. The colonial authorities had suggested that instead of exiling them offshore, a mainland reserve could be established so they could continue their traditional way of life; however, Robinson successfully argued against it. He deliberately opted to sever their spiritual connection to the land in order to prepare them to accept the Christian God.[19]

---

* At the time, company employees allegedly also committed numerous massacres of Aboriginal people (Pybus, 2020).

The island was too small to sustain them living traditionally and independently. There wasn't enough wild food, so they had to subsist on meagre handouts in abysmal conditions. Disease and death were rife. They were soon moved again to a settlement named Wybalenna on nearby Flinders Island. There was no reason to call it an improvement – they were exposed to the winds of the Roaring Forties, there were no freshwater springs or streams, and the wells they dug for water were brackish. Tasmania's original people continued to die there at an astonishing rate.

From 1833, what had started as a largely voluntary translocation became enforced, as Robinson rounded up roughly 300 people from across Tasmania at gunpoint. Many or most died in his so-called care before they even reached Wybalenna, while they were being dragged across the island or kept in holding camps or prisons.

By February 1835 Robinson declared, 'The entire Aboriginal population are now removed,'[20] and he was rewarded handsomely in cash payments and land grants by both the government and through a subscription arranged by the grateful public (although at least one family of Tarkiner people remained free on mainland Tasmania until their eventual capture in 1842).

As David Quammen puts it in *The Song of the Dodo* – the book that, when I read it as a teenager, became a massive influence on my becoming a zoologist – the conditions and the susceptibility to European diseases ensured that, 'In the mid-1840s . . . the death rate was higher than the birth rate. Precious few children survived infancy, and most of those were taken away to the Orphan School at Hobart, where their cultural identity was erased'.[21] It was no safer there, as at one point half the children at the Orphan School were killed by a measles epidemic.[22] Eventually, having realised the trauma caused by removing children from their parents, Robinson had them transported back to the squalor of Flinders Island, but they were forced to live with the camp's European overseers rather than their own families.

The Indigenous peoples' numbers became so small that it was no longer worth the government maintaining offshore Wybalenna. A number of the group's leading men organised for a petition to be sent to Queen Victoria arguing against the way they were being kept, having been dispossessed of their rightful lands. Some combination of those factors led to the forty-seven surviving people being moved to a derelict prison camp in Oyster Cove, near Hobart, in 1847, which had been abandoned as it was considered too unhealthy for convicts. The Indigenous Tasmanians were effectively left to die there in poverty, while the camp's overseers acted as pimps, trading sex with the female inhabitants for alcohol with nearby settlers.

A Nuenonne woman called Truganini – who had travelled with Robinson on his 'missions' as a translator, guide and essentially as an ambassador for her people – has often been described as 'the last Aboriginal Tasmanian', and it has been asserted that with her death in 1876, they became extinct. However, while the atrocious act of ethnic cleansing by the British was almost absolute, Indigenous Tasmanians did not die out when Truganini passed away. A number of people remained, chiefly those whose mothers had originally been kidnapped by their fathers – whalers, sealers or other settlers living on islands in the Bass Strait and on Kangaroo Island in South Australia. These women often lived lives of brutal torture, rape and forced labour, and were traded between the men for sealskins. They were largely ignored by the colonial infrastructure. Perhaps it was more convenient to think of them as extinct, as that would make reparation and reconciliation unnecessary. Nonetheless, their descendants now number in the thousands across the state. Aboriginal Tasmanians are not extinct.

Referencing the ongoing reports of thylacine sightings today, Tasmanian Aboriginal artist Vicki West wrote in 2014:

> The history of the 'extinction' of the thylacine closely parallels the 'extinction myth' of the Tasmanian Aboriginal people, with both having a bounty on their heads under colonial

rule, and with both being seen as pests in their own environment. The impact of this 'cleansing' of the landscape to conform to colonial pastoral ideals continues to be felt today.[23]

A caption in the excellent 'Our land: *parrawa, parrawa*! Go Away!' exhibition at the Tasmanian Museum and Art Gallery, which tells the story of the invasion from both sides, and was curated by members of the Tasmanian Aboriginal community, reads, 'We were never defeated, but we could not win.'

Indigenous Tasmanians and thylacines were both systematically targeted by European settlers and concerted efforts were made for their extirpation, including through government-sponsored bounties (although only the thylacine was completely lost). But the comparison does not end there. As the people's numbers declined, demand for their remains increased among museums. Graves were robbed and bodies mutilated to supply imperial collections such as those at the Royal College of Surgeons in London, Oxford University and the Royal Society of Tasmania. Truganini was buried after her death (within the walls of Hobart's female convict factory), but two years later, the Royal Society of Tasmania secretly dug her up and prepared her body for exhibition – something she had expressly said she didn't want to happen. Initially, her skeleton became a travelling exhibit and was then placed on display in the Tasmanian Museum from 1904 until 1947.* Casts of her skull and death mask were traded with museums around the world (the American Museum of Natural History swapped one for a *Tyrannosaurus rex* skull).[24] In a similar way, thylacine specimens

---

* Her skeleton was eventually repatriated to the Tasmanian Aboriginal community in 1976, the centenary of her death. She was cremated and her ashes scattered in the waters off Bruny Island, the ancestral home of the Nuenonne people. Hers were the first of now thousands of Aboriginal remains that museums have repatriated worldwide. In 2002, samples of Truganini's skin and hair were repatriated from London's Royal College of Surgeons, along with remains from four or five other Aboriginal people.

became much in demand by museums, as curiosities from the colonies. We have thylacine skins in Cambridge sent by the Hobart solicitor Morton Allport – the same man who supplied Tasmanian people's skulls to Oxford.

*

The marginalisation and 'replacement' of Indigenous society by the West was often characterised as an inevitable consequence of their alleged inferiority.[25] And this view carries over into how changes to the fauna have been portrayed post-colonisation. It is extremely common to hear it suggested that the reason that Australian animals have fared so poorly since 1788 is down to the inherent inferiority of marsupials in the face of new competition with the colonists' European animals: it was evolutionarily inevitable that the changes wrought by the colonists would result in their demise, albeit regrettably. The story goes that Australia's 'primitive' species had only managed to survive up until that point because they were cut off from encountering more 'advanced' animals from other parts of the world. And that obviously changed when Europeans arrived.

Indeed, the fact that Europeans and their animals have thrived in Australia at the expense of Aboriginal people and indigenous mammals could be used to support the narrative of Western superiority. Following the arrival of America's first live thylacines in 1902, Frank Baker, the superintendent of the US National Zoo, wrote in *The Washington Post*, 'Australian animals . . . have come to be the stupidest animals in the world, which accounts for their rapid extermination when the country was settled by the British . . . the thylacine belongs to a race of natural born idiots'.[26]

This also has the added benefit of placing the blame on the victims for failing as a result of their supposed shortcomings, which exonerates the colonists. However, this fails to take account of three key lines of evidence, which refute such a belief. First, Europeans introduced arguably *the most* effective and adaptable predators on the planet – cats and foxes (particularly when there are few other mammalian predators around anymore to keep them in check). These aren't just any old placental mammals – it is unfair to compare all placentals to all marsupials based on the exploits of these two species. It would be like concluding that one country was better than another at football, based on a game in which one country's most skilful professional players were pitched against eleven people selected at random from another country.

Second, it wasn't just the introduction of placental mammals that caused the extinctions and declines: there has been continental-scale habitat destruction. What species could reasonably be expected to survive that? To cut down a koala's tree and then blame it for dying is nothing but ludicrous.

And third, more than half the extinctions since Europeans arrived have been native placental mammals – rodents and bats – giving the lie to the claim that it's because of their marsupiality that species haven't been able to cope.

What's more, the idea that marsupials only managed to survive in Australia because they were isolated and without competition from placentals is also undermined by the fact that marsupials have lived and do live alongside placental mammals, and not just rodents and bats. As I mentioned, there are around 100 marsupial species in the Americas showing no signs of being outcompeted into oblivion by the placentals around them. And in Australia there is an extremely enigmatic single fossil tooth – from the same site where the world's oldest songbirds have been found – which appears to be from a 55-million-year-old placental mammal called *Tingamarra*.[27] This tears up the story that marsupials would have been outcompeted if only placentals had made their way to Australia earlier on. Placentals *did* arrive, but apparently failed to cope as well as the marsupials around them, and went extinct.

I've spent most of this book attempting to demonstrate that despite what you might have heard, there is nothing inferior about Australian mammals. My point here is that the same is true of its people. The idea that Australia is inherently inferior – be that a largely subconscious view that has tainted impressions of its animals and its people – is propaganda that has helped to oil the colonial machine.

*

It's curious that Australian wildlife is belittled in the ways I've described, as arguably this does not apply to other parts of the former British Empire, like India. The expansion of imperial frontiers in Africa and India was facilitated by the opening up of lands for big-game hunting. Moustachioed toffs in pith helmets travelled to those places specifically to murder mammals, in order to nail them to their walls as supposed demonstrations of their epic manliness. My agent, Doug Young, suggested to me that this afforded Indian mammals, for example – like tigers, rhinos and elephants – a higher societal value, on account of them being majestic enough to want to turn into trophies, whereas 'you couldn't say the same about a wombat'.

I've decided to save an argument about the majesty of wombats for another day, but it's certainly true that the hunting safari industry didn't exist in colonial-era Australia on any level remotely comparable to India or Africa. In fact, red foxes and deer were deliberately introduced to Australia specifically so that the colonists could hunt them for sport. This implies that, despite the abundance of wildlife they originally found there, hunters were underwhelmed by the opportunities to kill things offered by the native wildlife.

With that said, kangaroo hunting was a significant part of life for settlers in the nineteenth century. This started as a necessity for survival – Europeans initially hunted kangaroos primarily for meat. As I've discussed, in early colonial Tasmania, this was one of the key factors driving conflict between Indigenous populations and the colonisers, as it starved Aboriginal Tasmanians of one of their principal protein sources.

In terms of the *terra nullius* argument, the irony is that it's clear from many accounts that the new settlers were reliant on kangaroo hunting for food. So, who really were the hunter-gatherers? Seemingly the white settlers included it in their fledgling societies as much as the Indigenous populations. Furthermore, hunting native animals is in itself an act of colonisation – it formed part of the Europeans' claim on the land; whereas when it was Aboriginal people doing the hunting, the settlers used it as an argument that, as hunter-gatherers, they didn't own the land. It was a complete double standard.

In their 2020 book, *The Colonial Kangaroo Hunt*, Ken Gelder and Rachael Weaver explain how in 1830s Tasmania, the Black War tensions over kangaroo hunting eased after the Aboriginal people had largely been driven offshore, paving the way for hunting as entertainment to take hold. This mirrored a similar increase in sport hunting in New South Wales and Victoria. Effectively, hunting for fun, rather than food, was a 'post-frontier' activity – an act of civilisation, after the violence of the frontier, 'providing settlers with a means to enjoy the land they had violently claimed and occupied', and 'an expression of settler domination over species and territory'.[28]

All in all, kangaroo hunts were modelled far more on British foxhunting – complete with horses, hounds, horns and scarlet jackets – than they were on big-game hunting safaris. The quality of the hunting was used as propaganda to lure people from Britain to emigrate to Australia, but it was a distinctly local pastime. While visiting dignitaries were occasionally taken on organised kangaroo hunts, it was nothing on the scale of the expeditions that colonial hunters mounted in Africa and South Asia, and it was only for one kind of animal.

One parallel that does exist between hunting kangaroos and the safari industry elsewhere is the role they both played in expanding the limits of the colonial frontier. The surveyor George William Evans demonstrated that hunting kangaroos and emus required 'sportsmen' to open up new regions of Australia to the Europeans: 'They are both timid in their nature, and soon abandon districts which are settled, retiring back into the cultivated wilds. To enjoy these sports, therefore, in perfection, it is necessary to go beyond the limits of colonization'.[29]

Despite all this hunting, I can't remember ever having seen a mounted taxidermy trophy head of a native Australian animal (although you certainly find European-style deer heads in Australian 'wilderness hotels' aspiring to a certain country retreat vibe).* Again, this demonstrates a difference between the kind of colonial hunting that was taking place in Australia and other parts of the Empire.

Sadly, trophy hunting continues today, particularly in Africa and North America, and in 2019 it was uncovered that a casino in Victoria was advertising opportunities to high-rolling Chinese clients to stay at a luxury hunting lodge and go out and shoot local wombats,[30] so it seems these majestic – I said MAJESTIC – marsupials have not in fact been spared this senseless slaughter. In my mind, anyone who thinks that killing animals is fun is fundamentally wrong, but having stalked hundreds of wombats for far less murderous reasons, I can tell you that there would be about as much sport in shooting at wombats as there is in shooting at a bale of hay.

---

*There are two thylacine skulls on wooden backboards, with jaws agape, at the National Museum of Ireland – Natural History that could fit this description, though they are not taxidermy. Like lions, tigers and wolves – which also suffer the indignation of being regularly mounted as trophy heads – thylacines had a reputation for being ferocious. While this was also propaganda that helped justify their extirpation, it would explain why this species might be mounted in this way, perhaps uniquely among Australian wildlife, as a beast a boastful hunter would want to demonstrate that they had overcome.

Doug's suggestion is that hunters considered Asian and African big game mammals to be 'noble', unlike the Australian fauna, and I think there's a lot to be said for that. The absence of 'big game' in Australia was likely a factor in the differences between attitudes to Indian and Australian wildlife, but I think it's also because, to a large extent, the animals of Asia merge seamlessly with the animals of Europe, and had been part of the 'known world' for centuries.

Comparisons with the Americas would be more instructive (contact was made anew, following massive sea voyages, with no long-term historical knowledge of its peoples), but again the animals that settlers and explorers found there were mainly similar to or the same as the ones they had found in Eurasia (because they are periodically physically connected via a land bridge between what is now Alaska and Russia).

But with that said, there *was* a tendency to denigrate the animals of the Americas at a similar point in history. Georges-Louis Leclerc, Comte de Buffon, was a French aristocrat and one of the most prominent scientists of the eighteenth century. Following in the footsteps of a number of European writers, Buffon asserted that American animals were altogether inferior to their European counterparts. This is despite the fact that he had never actually crossed the Atlantic to see them for himself.

Unlike Australia, where the animals would be considered entirely different to those in the 'Old World', Buffon made explicit reference to the fact that both European and American animals were of the same types, but he insisted that those in the Americas were always smaller and weaker.[31] It doesn't require too much consideration to see how these claims don't stand up to scrutiny (bison and grizzly bears are massive, for one thing).

As with Australia, we can draw parallels between how Europeans described America's animals and how they described its Indigenous peoples. As well as writing off its mammals, Buffon said this:

> In the savage the organs of generation are small and feeble; he has no hair, no beard, no ardour for the female; though more nimble than the European, from being habituated to running, he is not so strong; possessed of less sensibility, yet he is more timid and dastardly; he has no vivacity, no activity of soul . . . Satisfy his hunger and thirst and you annihilate the active principle of all his motions; and he will remain for days together in a state of stupid inactivity.[32]

We should probably pay little consideration to a description that puts penis size at the top of a list of comparators.

It's also interesting to note that other eighteenth-century writers used the Americas' complement of dangerous snakes and invertebrates as evidence of its degeneracy, as I suggest people do subconsciously regarding Australia today.

All of this compares closely with Australia's own treatment at the hands of European scientists at the time, and once again we don't have to look far for potential political benefits from undermining the distant continent. Buffon was writing around the time of the American Revolution. Although France supported America as it fought for independence from Great Britain, the burgeoning democracy exemplified in the United States undermined the European structures of colonialism upon which it was originally founded. Not to mention that all of this shortly preceded the French Revolution. America was not a model that Europe wanted to celebrate, and disparaging its wildlife and its people was one tactic to paint it as inferior.

Where the comparisons end, however, is that these early attacks on the 'quality' of American wildlife have not stood the test of time, unlike in Australia. Although they are not universally loved (particularly by US pastoralists), I think that the general opinion of America's iconic wildlife – bison, wolves, moose, eagles, bears and pumas, for example – falls into the 'noble and majestic' camp. And it may not escape your notice that these are all considered desirable by trophy hunters.

*

Since 1800, when Johann Blumenbach attempted to name them *Ornithorhynchus paradoxus*, platypuses have been considered a paradox. Perhaps the argument running through this book also appears paradoxical. Unquestionably, Australia adores its wildlife, and that sentiment is shared by people the world over. Platypuses are popular, and on the face of it, the ways that Australian mammals are described today are largely intended to be positive. Few, if any, countries are so intimately associated with an animal as Australia is with kangaroos, or koalas, or platypuses. And so, is it fair to say that they are negatively portrayed?

No doubt, it is hard to look at most Australian mammals and have any emotion other than joy. But that doesn't mean that we aren't subconsciously making other judgments, too. Examine any description of Australian wildlife and chances are, alongside positive words like 'wonderful' and 'unique', we will spot pejorative little jibes implying they are weird or bizarre oddities. My ears have become sensitive to it. I spend a lot of time talking to people about Aussie mammals, and I think that I am generally successful at stoking enthusiasm and wonder for them. Nevertheless, at the end of these conversations, the other person almost always says, 'That's so *strange!*' They don't mean anything by it, but I wonder where it comes from. I think we have somehow been conditioned to trot out these words when given to think about these species.

The only conclusion I can reach is that we have been influenced by the centuries of historical baggage attached to perceptions of Australia and its wildlife. Over time, the scientific interpretations of the world – and the political and societal outcomes of those interpretations – have both influenced and been influenced by attitudes to Australian mammals. I doubt that the architects of the British Empire sat around strategising that by denigrating the country's fauna, they could maximise their gains from the Australian colony at the expense of its Traditional Owners. Nor do I think that industrialists today have deliberately contrived to make conservation difficult by making the world assume its mammals are inferior. Nonetheless, gains were made, and conservation is difficult. And I can't help but think that things might be different if greater value had been assigned to platypuses and possums. In this way, it really does matter how we talk about them. It has always mattered.

Afterword: A Call to Arms

We need to change the way we talk about Australian animals, because the current narrative perpetuates a colonial view of the country and its peoples, and hinders environmental conservation. We need to resist the notion of 'weird' animals, because this is a belittling value judgment. We need to stop calling animals primitive, because it doesn't make any scientific sense.

When we go into a museum that implies in its displays that Australian animals are anything but precisely adapted results of evolution, we need to call them out and explain the harm they could be doing by conveying messages that work against what should be their primary mission – engaging people with the natural world.

Museums themselves also need to think about the fact that their specimens may not actually look like the animals they are representing, and consider why that might be problematic. Is it better to teach people the wrong thing, or nothing at all? That's not an easy question to answer.

TV and film producers need to back away from the lazy trope that Australia's wildlife makes it a dangerous place, or that its animals are comically dopey. Why not instead explore the narrative that it is the only place on Earth where all three groups of mammals are found, and that its extraordinary diversity is disappearing before our very eyes?

And us . . . we need to think about where our attitudes and

assumptions have come from. What are we really saying when we call animals 'bizarre' or 'weird'?

Platypuses, possums, wombats, echidnas, devils, kangaroos, quolls, dibblers, dunnarts, kowaris – and many of the other Australians few people have ever heard of – are *wonderful*, and they deserve our respect.

Acknowledgements

A year after I had settled into my new job in Cambridge, in the very week that I had privately decided I had the mental space to start writing my next book, about platypuses, I received an email from Doug Young. By startling coincidence, Doug said he had read my previous book, and based on its first chapter he was wondering whether I would be interested in writing a book about platypuses, and if he could represent me as my agent.

Along with Doug and Pew Literary, my editors – Myles Archibald and Hazel Eriksson at HarperCollins and Joseph Calamia at the University of Chicago Press – have been wonderful in helping me piece this all together coherently and to make the book better. I am extremely grateful for all their work and advice.

The thoughts laid out in this book are the result of countless interactions with people, animals and places over the last twenty years, and of learning from innumerable writers and naturalists who have come before. It would be impossible to thank them all.

Those ecologists who lead the field trips I take part in have undreamed-of impacts on my life, through building both experience and knowledge, but also through the deep friendships that the intimacy of remote fieldwork fosters. At risk of missing someone out, I cannot sufficiently thank Chris Dickman,

Aaron Greenville, Rodrigo Hamede Ross, David Hamilton, Rosie Hohnen, Tracey Hollings, Chantelle Jackson, Alex James, Hugh McGregor, Céline Mazier, Andrew Morton, Bryony Palmer, James Smith, Katherine Tuft, Georgia Ward-Fear, Rohan Wilson and Karen Young, as well as the University of Tasmania's devil research programme, the Australian Wildlife Conservancy and Arid Recovery. In addition, adventures with Lyndal and Leigh Byford, Joey Clarke, Andrew Foster and Toby Nowlan have introduced me to many Australian mammals.

My passion for platypuses was kick-started by my lecturer (and now colleague) Adrian Friday, in his undergraduate class in Cambridge. Those classes also introduced me to museums, and I am so very grateful to my colleagues who became family after fourteen years at the Grant Museum of Zoology at UCL, and now to the wonderful team at the University Museum of Zoology, Cambridge.

A number of people shared their ideas with me through conversations about this book. Thanks to: Subhadra Das for thoughts on eugenics; Alison Douglas for taxidermy echidnas; Rosemary Fleay-Thomson for her father's platypuses; Kristofer Helgen and Roberto Portela Miguez for echidna chats; Naomi Indigo for quoll sausages; Ayesha Keshani for colonial specimen preparation; Matt Lowe for Caldwell's embryos; George Madani for his experiences at the jaws of a loris; Bethany Palumbo for thylacine stripes; Katrina Schlunke for thylacine 'primitivity'; Stephen Sleightholme for his thylacine suggestions and generous sharing of materials (and for producing the International Thylacine Specimen Database); Jana Stewart for our time spent searching for Tasmanian platypuses and for sharing research papers; and Laura Wills at the Literary and Philosophical Society in Newcastle and Dan Gordon at the Great North Museum for answering questions about their platypus specimen. Thanks to Natalie Jones for photographing our specimens in Cambridge.

I often marvel at how much easier researching natural history books has become since the creation of the Biodiversity Heritage Library – it is amazing, and I am grateful for its existence.

Katherine Tuft, Kristofer Helgen and Georgia Ward-Fear were extremely generous with their time and expertise, by reading a full draft and providing encouragement and invaluable suggestions – the book is improved as a result.

Writing in the bizarre isolating conditions of the 2020 pandemic has been a curious experience. I must say a huge thank you to the friends who kept me sane: Mark Carnall, Helen Chatterjee, Hannah Cornish, Subhadra Das, Tannis Davidson, Sarah Dellar, Ruth Desforges, Jayne Dunn, Andrew Foster, Pete Hinstridge, Clem Liebenberg, Erica McAlister, Sebastien Penigault, Rona Strawbridge, An Van Camp, Dean Veall, Chris Wearden, Steve Westlake, Shaun Whelan and the NatSCA gang.

And finally, thank you to my family, who are wonderful: my parents, Sally, Gavin, Alfie, Sadie and the endless energy, silliness and support of my Big Family.

About the Author

Jack Ashby is the assistant director of the University Museum of Zoology, Cambridge, one of the UK's largest and most significant natural history museums. His life is split between a career dedicated to engaging people with the natural world – chiefly through museums – and ecological fieldwork across Australia, on behalf of universities and wildlife organisations there. He is the author of *Animal Kingdom: A Natural History in 100 Objects* (The History Press, 2017), which explores what we can learn about the incredible processes behind evolution from specimens in museums, as well as the sometimes unscientific, political ways in which natural history museums provide an unnatural view of nature.

Jack is a trustee of the Natural Sciences Collections Association (the UK's professional body for museums with natural history collections) and formerly of the international Society for the History of Natural History. He is also an honorary research fellow in the Department of Science and Technology Studies at University College London; and is currently undertaking a Headley Fellowship, supported by Art Fund, researching the colonial history of the Australian mammal collections in Cambridge.

When he's not chasing animals for work, he's generally doing it for fun, undertaking mammal-watching trips around the world, and sharing his excitement for what he finds on social media. He lives in Hertfordshire.

Endnotes

Preface

1 Bethge et al., 2001; Bethge, 2002

Introduction

1 Fox, 1827
2 Hunter, 1798a
3 Hunter, 1798b
4 Banks, 1798
5 Home, 1802
6 Collins, 1802
7 Burrell, 1927
8 Lewis, 1996
9 Owen, 1834c
10 Weisbecker & Goswami, 2010
11 Burrell, 1927
12 Parker, 1885
13 Gould, 1863
14 IUCN, 2020

1. Meet the Platypus

1 Anich et al., 2021
2 Manger & Pettigrew, 1995
3 Burrell, 1927
4 Bennett, 1835
5 Fleay, 1980
6 Burrell, 1927
7 Burrell, 1927
8 Burrell, 1927
9 Burrell, 1927
10 Bennett, 1835
11 Burrell, 1927
12 Bethge, 2002
13 Grant, 1984
14 Thomas et al., 2018a
15 Thomas et al., 2018a
16 Hughes & Hall, 1998
17 Thomas et al., 2018b
18 Manger et al., 1998
19 Manger et al., 1998
20 Grant, 1984
21 Wood, 1865
22 Jackson, 2002
23 Manger et al., 1998
24 Thomas et al., 2018b
25 Burrell, 1927
26 Jamison, 1818
27 Fleay, 1980
28 Burrell, 1927
29 Whittington & Belov, 2007
30 Koh et al., 2007
31 Hurum et al., 2006
32 Enjapoori et al., 2014
33 Friday, 2014

34 Bethge et al., 2001
35 Flannery et al., 2022a
36 Weisbecker & Beck, 2015; Chimento et al., 2023
37 Pian et al., 2013
38 Bino et al., 2019
39 Paterson, 1933
40 Burgin et al., 2018
41 Vergnani, 2019
42 Sullivan, 2020; Smith, 2020

2. Diplomatic Platypuses

1 Cushing & Markwell, 2009
2 Fleay, 1947
3 Lawrence, 2017
4 San Diego Zoo website, Accessed April 14 2020
5 Burrell, 1927
6 Burrell, 1927
7 Burrell, 1927
8 Fleay-Thomson, 2007
9 Fleay-Thomson, 2007
10 Anonymous, 1944
11 Fleay, 1980
12 Thomas, 2018
13 Fleay, 1980
14 Fleay-Thomson, 2007
15 Fleay, 1980
16 Australian Museum website, Accessed 16 April 2020
17 Cushing & Markwell, 2009
18 Fleay-Thomson, 2007
19 Anonymous, 1947
20 Bernheim, 1947
21 Fleay, 1947
22 Fleay-Thomson, 2007
23 Fleay-Thomson, 2007
24 Australia. Parliament. Senate. Rural and Regional Affairs and Transport References Committee, 1998
25 Fikes & Wilkens, 2019

3. Echidnas: The Other 'Primitives'

1 Wallace, 1876
2 Shaw, 1792
3 Griffiths, 1978
4 Lord, 1920
5 Olsen, 2010
6 Griffiths, 1978
7 Cabrera, 1919
8 Helgen et al., 2012
9 Nicol, 2015
10 Nicol, 2017
11 Bino et al., 2019
12 Clemente et al., 2016
13 Nicol, 2015
14 Burrell, 1927
15 Nicol, 2015
16 Morrow et al., 2009
17 Johnston et al., 2007
18 Fenelon, 2021
19 Johnston et al., 2007
20 Owen, 1865
21 Nicol, 2015
22 Summerell et al., 2019
23 Cormack, 2015
24 Augee, 1976

4. The Mystery of the Egg-laying Mammals

1 Shaw, 1799
2 Shaw, 1800
3 Darwin, 1839
4 Burrell, 1927
5 Blumenbach, 1797
6 Owen, 1832

7 Burrell, 1927
8 Mahoney, 1988
9 Home, 1802
10 Blumenbach, 1800
11 Home, 1802
12 Olsen & Russell, 2019
13 Bennett, 1835
14 Pybus, 2020
15 Unaipon, 1924–1925
16 Unaipon, 1924–1925
17 McKay et al., 2001
18 McKay et al., 2001
19 Olsen & Russell, 2019
20 Olsen & Russell, 2019
21 Olsen & Russell, 2019
22 Olsen & Russell, 2019
23 Burrell, 1927
24 Olsen & Russell, 2019
25 Darwin, 1872a
26 Darwin, 1860
27 Owen, 1832
28 Owen, 1832
29 Zoological Society of London, 1832
30 Owen, 1834a
31 Owen, 1834a
32 Hall, 1999
33 Owen, 1865
34 Owen, 1865
35 Nicol, 2018

5. Cracked

1 Friday, 2016
2 Bidder, 1941
3 Science Museum Group Collection Online, 1884
4 Caldwell, 1884
5 Olsen & Russell, 2019
6 Caldwell, 1887
7 Caldwell, 1884

8 Moseley, 1885
9 Haacke, 1885
10 Haacke, 1885
11 Anonymous, 1885
12 Olsen & Russell, 2019
13 Caldwell, 1884
14 Caldwell, 1884
15 Caldwell, 1884
16 Caldwell, 1887
17 Caldwell, 1887
18 Caldwell, 1887
19 Olsen & Russell, 2019
20 Caldwell, 1887
21 Burrell, 1927
22 Moyal, 1986
23 Field, 1825
24 Jamison, 1818
25 Nicol, 2018
26 Owen, 1832
27 Moyal, 2001

6. Terrible with Names

1 Bennett, 1835
2 Lydekker, 1896
3 Owen, 1834c
4 Ashby, 2017b
5 Anonymous, 1864
6 Ashby, 2018
7 Triggs, 1988
8 Sample, 2018
9 Magondu et al., 2023
10 Triggs, 1988
11 Triggs, 1988
12 Smith, 1972
13 Parrish, 1997
14 Whitley, 1970
15 McHugh, 2006
16 McHugh, 2006
17 Whitley, 1970
18 Simons, 2013
19 George, 1964

20 McHugh, 2006
21 Von Husyett, 1994
22 Whitley, 1970
23 Jackson & Vernes, 2010
24 Dampier, 1697, pub. 2007
25 Dampier, 1703, pub. 1961
26 Ashby, 2015
27 Simons, 2013
28 Ashby, 2013
29 McHugh, 2006
30 Ashby, 2014
31 Cash, c.1905
32 Banks, 1770
33 Ashby, 2012
34 Banks, 1770
35 Tench, 1793
36 Fender-Barnett, 2019
37 Low, 2016

7. Marvelling at Marsupials

1 Strachey, 1612, pub. 1849
2 Hugghins & Potter, 1959
3 Fleay, 1980
4 Wood, 1865
5 Cremona et al., 2021
6 Jackson, 2015
7 Van Dyck & Strahan, 2008
8 Wooller, 2015
9 Baker, 2015
10 Wroe et al., 2005
11 Arlington, 2016
12 Cuvier, 1827
13 Hohnen et al., 2012
14 Van Dyck & Strahan, 2008
15 Elder et al., 2015
16 Weisbecker & Nilsson, 2008
17 Jackson & Vernes, 2010
18 Eldridge & Coulson, 2015
19 Dampier, 1703, pub. 1961
20 Van Dyck & Strahan, 2008

21 Finlayson, 1935
22 Gould, 1863
23 Burgin et al., 2018
24 Van Dyck & Strahan, 2008
25 Finlayson, 1935
26 Van Dyck & Strahan, 2008

8. Roo-production

1 Bennett & Goswami, 2011
2 Travouillon et al., 2019
3 Roberts & Schulz, 2016
4 Shaw, 2006
5 Wooller, 2015
6 Gould, 1863
7 Gelder & Weaver, 2020
8 Menzies et al., 2020
9 Baker, 2015
10 Chan et al., 2020
11 Van Dyck & Strahan, 2008

9. The Missing Marsupials

1 Johnson, 2006
2 Woodford, 2002
3 Long et al., 2002
4 Long et al., 2002
5 Owen, 1859
6 Wroe et al., 2005
7 Woodward, 1890
8 Akerman & Willing, 2009
9 Pickrell, 2018
10 Johnson, 2006
11 Wroe et al., 2013

10. Copycats and Cover Versions

1 Ashby, 2017a
2 Sleightholme & Ayliffe, 2013
3 Feigin et al., 2018

4 Darwin, 1872b
5 Paddle, 2000
6 Owen, 1834c
7 Harris, 1808
8 Newton et al., 2018
9 Freeman, 2014
10 Harris, 1808
11 Gelder & Weaver, 2020
12 Freeman, 2014
13 Freeman, 2014
14 Wood, 1865
15 Paddle, 2000

11. They're Stuffed

1 Carnall, 2013
2 Shaw, 1792
3 Ashby, 2020a
4 Bewick, 1800
5 Bewick, 1800
6 Pigott & Jessop, 2007
7 Cooper et al., 2019
8 Sleightholme & Ayliffe, 2013
9 Van Wyhe & Drawhorn, 2015
10 Ashby, 2020b
11 Van Wyhe & Drawhorn, 2015
12 Wallace, 1869
13 Olsen & Russell, 2019

12. Extinction

1 Woinarski et al., 2015
2 Legge et al., 2018
3 Wellauer & Thomas, 2020
4 Cogger et al., 2003
5 Gammage, 2011
6 Radford et al., 2018
7 Legge et al., 2020
8 Legge et al., 2020
9 Australian Government, 2015
10 Department of the Environment

and Energy, 2019
11 Doherty et al., 2019
12 Legge et al., 2020
13 Legge et al., 2020
14 Shine et al., 2020
15 Ward-Fear et al., 2017
16 Indigo et al., 2018
17 Decker et al., 2019
18 Mills, 2016
19 Travouillon et al., 2019
20 Travouillon & Phillips, 2018
21 Cox, 2018
22 Tuft, 2018
23 Lankester, 1885
24 Bino et al., 2019
25 Haygarth, 2017
26 Bino et al., 2019
27 Grant, 1984
28 Burrell, 1927
29 Grant, 1984
30 Woinarski & Burbidge, 2016
31 Bino et al., 2019
32 Klamt & Davis, 2011
33 Victoria State Government, 2021
34 Australian Conservation Foundation, 2020
35 Samuel, 2020
36 Cox, 2020a
37 Cox, 2020b
38 Slezak, 2022

13. *Terra Nullius* and Colonialism

1 Boldrewood, 1901
2 Bond, 2017
3 Eldridge & Browning, 2002
4 Behrendt, 2016
5 Pascoe, 2018
6 Gammage, 2011

7   Gammage, 2011
8   Gammage, 2011
9   Gammage, 2011
10 Dampier, 1697, pub. 2007
11 Gammage, 2011
12 Pascoe, 2018
13 Pascoe, 2018
14 Gammage, 2011
15 Sutton and Walshe, 2021
16 Gammage, 2011
17 Wahlquist, 2020
18 Borschmann et al., 2020
19 Pybus, 2020

20 Quammen, 1996
21 Quammen, 1996
22 Pybus, 2020
23 Maynard & Gordon, 2014
24 Pybus, 2020
25 Wood, 1865
26 Paddle, 2000
27 Long et al., 2002
28 Gelder & Weaver, 2020
29 Evans, 1822
30 Burke, 2019
31 Thomson, 2008
32 Buffon, 1807

References

Akerman, K. and Willing, T., 2009. An ancient rock painting of a marsupial lion, *Thylacoleo carnifex*, from the Kimberley, Western Australia. *Antiquity,* 83(319), pp. 1–4

Anich, P., Anthony, S., Carlson, M., Gunnelson, A., Kohler, A., Martin, J. and Olson, E., 2021. Biofluorescence in the platypus (*Ornithorhynchus anatinus*). *Mammalia,* 85(2), pp. 179–181

Anonymous, 1864. Phascolomes. *Le Magasin Pittoresque,* pp. 28–29

Anonymous, 1885. Societies and academies: Royal Society of New South Wales, meeting report from November 5th 1884. *Nature,* 31, pp. 235–236

Anonymous, 1944. First duck-bill platypus born in captivity. *The Illustrated London News,* 25 March, p. 19

Anonymous, 1947. The incredible platypus. *The New York Times,* 30 April, p. 24

Arlington, K., 2016. The eastern quoll, thought extinct on mainland for 50 years, may yet be alive. *The Age,* 26 February. Accessed online 28 September 2020

Ashby, J., 2012. Kangaroos cooked up by Cook. *UCL Museums Blog,* 13 March. Accessed online 16 May 2020

Ashby, J., 2013. It's Australia v England, in battle over Stubbs masterpieces. *The Conversation,* 7 November. Accessed online 16 May 2020

Ashby, J., 2014. The kangaroo painting that might never have been – following in Cook's footsteps. *Travellers' Tails, Royal Museums Greenwich*, 23 October. Accessed online 16 May 2020

Ashby, J., 2015. The earliest Strange Creatures: Europe's first meetings with marsupials. *UCL Museums Blog*, 2015 May. Accessed online 16 May 2015

Ashby, J., 2017a. *Animal Kingdom: A Natural History in 100 Objects.* Gloucester: The History Press

Ashby, J., 2017b. Does an animal's name affect whether people care about it?. *Arid Recovery Blog*, 8 September. Accessed online 16 May 2020

Ashby, J., 2018. The taxidermy koala – the language of natural history. *UCL Museums Blog*, 16 February. Accessed online 16 May 2020

Ashby, J., 2020a. Tasting platypus milk: linking specimens and stories. *Biodiversity Heritage Library Blog*, 7 July. Accessed online 9 July 2020

Ashby, J., 2020b. Telling the truth about who really collected the 'hero collections'. *The Natural Sciences Collections Association Blog*, 22 October. Accessed online 7 November 2020

Augee, M., 1976. Heat tolerance of monotremes. *Journal of Thermal Biology,* 1(3), pp. 181–184

Australia. Parliament. Senate. Rural and Regional Affairs and Transport References Committee, 1998. *Commercial utilisation of Australian native wildlife, The report of the Senate Rural and Regional Affairs and Transport References Committee,* Canberra: The Committee

Australian Conservation Foundation, 2020. *Platypus in decline,* Carlton, Victoria: Australian Conservation Foundation

Australian Government, 2015. *Threatened species strategy*

Australian Museum website. Accessed 16 April 2020 *https://australianmuseum.net.au/about/history/exhibitions/trailblazers/william-bligh/*

Baker, A. M., 2015. Family Dasyuridae (carnivorous marsupials). In: D. Wilson & R. Mittermeier, eds. *Handbook of the Mammals of the World. vol. 5. Monotremes and Marsupials.* Barcelona: Lynx Edicions, pp. 232–349

Baker, K., Eichhorn, M. and Griffiths, M., 2019. Decolonizing field ecology. *Biotropica*, 51, pp. 288–292

Banks, J., 1770. *The Endeavour journal of Sir Joseph Banks.* Reproduced online: Project Gutenberg Australia

Banks, J., 1798. *Letter to The Literary and Philosophical Society of Newcastle upon Tyne, dated 24th September 1798.* Literary and Philosophical Society Library, Newcastle upon Tyne.

Barlow, N. (ed.), 1958. *The Autobiography of Charles Darwin 1809–1882.* London: Collins

Behrendt, L., 2016. Indigenous Australians know we're the oldest living culture – it's in our Dreamtime. *Guardian*, 22 September. Accessed online 25 May 2020

Bennett, C. and Goswami, A., 2011. Does reproductive strategy drive limb integration in marsupials and monotremes? *Mammal Biology,* 76, pp. 79–83

Bennett, G., 1835. Notes on the natural history and habits of the *Ornithorhynchus paradoxus*, Blum. *Transactions of the Zoological Society of London,* 1, pp. 229–258

Bernheim, B. M., 1947. Platypus I.Q. defended. *The New York Times*, 9 May, p. 20

Bethge, P., 2002. *Energetics and foraging behaviour of the platypus* Ornithorhynchus anatinus. PhD thesis: University of Tasmania

Bethge, P., Munks, S. and Nicol. S., 2001. Energetics of locomotion and foraging behaviour in the platypus *Ornithorhynchus anatinus. Journal of Comparative Physiology B,* 171(6), pp. 497–506

Bewick, T., 1800. *A General History of Quadrupeds.* Newcastle upon Tyne: S. Hodgson, R. Beilby & T. Bewick

Bidder, G., 1941. Mr W. H. Caldwell. *Nature,* 148, pp. 557–559

Bino, G., Kingsford, R. T., Archer, M. Connolly, J. H., Day, J., Dias, K., Goldney, D., Gongora, J., Grant, T., Griffiths,

J., Hawke, T., Klamt, M., Lunney, D.,Mijangos, L., Munks, S., Sherwin, W., Serena, M., Temple-Smith, P., Thomas, J., Williams, G., Whittington, C., 2019. The platypus: evolutionary history, biology, and an uncertain future. *Journal of Mammalogy,* 100(2), pp. 308–327

Blumenbach, J., 1797. *Handbuch der Naturgeschichte.* 4th ed. Göttingen: Johann Christian Dieterich

Blumenbach, J., 1800. Sur un nouveau genre de quadrupede édenté, nommé Ornithorhynchus paradoxus. *Bulletin Des Sciences, Par La Société Philomatique De Paris,* 2(39), p. 113

Boldrewood, R., 1901. *In Bad Company and Other Stories.* New York: Macmillan & Co

Bond, C., 2017. Fifty years on from the 1967 referendum, it's time to tell the truth about race. *The Conversation,* 30 May. Accessed online 25 May 2020

Borschmann, G., Gordon, O. and Mitchell, S., 2020. Rio Tinto blasting of 46,000-year-old Aboriginal sites compared to Islamic State's destruction in Palmyra. *ABC News,* 29 May. Accessed online 30 May 2020

Buffon, G. L. L. c. d., 1807. *Buffon's Natural History, Volume 7.* Translated ed. London: H.D. Symonds

Bullock, T., 1999. The future of research on electroreception and electrocommunication. *Journal of Experimental Biology,* 202, pp. 1455–1458

Burgin, C. J., Colella, J. P., Kahn, P. L. and Upham, N. S., 2018. How many species of mammals are there? *Journal of Mammalogy,* 99(1), pp. 1–14

Burke, K., 2019. Wombats 'being hunted by Chinese tourists at luxury Victorian resort', police investigate. *7News.com. au,* 1 August. Accessed online 28 May 2020

Burrell, H., 1927. *The Platypus.* Sydney: Angus and Robertson

Cabrera, A., 1919. *Genera Mammalium.* Madrid: Museo Nacional de Ciencias Naturales

Calaby, J. H., 1965. Early European description of an Australian mammal. *Nature,* 205, pp. 516–517

Caldwell, W. H., 1884. On the development of the monotremes and *Ceratodus. Journal of the Royal Society of New South Wales,* 18, pp. 117–122

Caldwell, W. H., 1887. The embryology of monotremata and marsupialia. Part I. *Philosophical Transactions of the Royal Society B,* 178, pp. 463–486

Carnall, M., 2013. Natural history museum bingo!. *UCL Museums Blog,* 15 October. Accessed online 4 December 2020

Cash, C. G. (ed), *c.*1905. *Cook's Voyages. The Life and Voyages of Captain James Cook. Selections with introductions and notes.* London: Blackie and Son

Chan, R., Dunlop, J. and Spencer, P., 2020. Highly promiscuous paternity in mainland and island populations of the endangered northern quoll. *Journal of Zoology,* 310, pp. 210–220

Chimento, N. R., Agnolín, F. L., Manabe, M. et al., 2023. First monotreme from the Late Cretaceous of South America, *Communications Biology,* 6, 146

Clemente, C. J. Cooper, C. E. Withers, P. C. Freakley, C., Singh, S. and Terrill, P., 2016. The private life of echidnas: using accelerometry and GPS to examine field biomechanics and assess the ecological impact of a widespread, semi-fossorial monotreme. *Journal of Experimental Biology,* 219, pp. 3271–3283

Cogger, H., Ford, H, Johnson, C. J., Holman, J. and Butler, D., 2003. *Impacts of land clearing on Australian wildlife in Queensland,* Brisbane: World Wide Fund for Nature Australia

Collins, D., 1802. *An account of the English colony in New South Wales, from its first settlement in January 1788 to August 1801.* London: Cadell & Davies

Connolly, J., Obendorf, D., Whittington, R. and Muir, D., 1998. Causes of morbidity and mortality in platypus (*Ornithorhynchus anatinus*) from Tasmania, with particular reference to Mucor amphibiorum infection. *Australian Mammalogy,* 20, pp. 177–187

Cooper, N., Bond, A. L. Davis, J. L. Portela Miguez, R., Tomsett, L. and Helgen, K. M. 2019. Sex biases in bird and mammal natural history collections. *Proceedings of the Royal Society B*, 289, 20192025

Cormack, L., 2015. New species of glider discovered in the Northern Territory. *The Sydney Morning Herald*, 12 August

Cox, L., 2018. Endangered bandicoot 'should never have been brought to South Australia'. *Guardian*, 17 July. Accessed online 15 August 2020

Cox, L., 2020a. Australia's environment laws: how do they work and what needs to be done to fix them? *Guardian*, 30 June. Accessed online 19 September 2020

Cox, L., 2020b. Australia's environment in unsustainable state of decline, major review finds. *Guardian*, 2020 July. Accessed online 19 September 2020

Cremona, T., Baker, A. M. Cooper, S. J. B., Montague-Drake, R., Stobo-Wilson, A. M. and Carthew, S. M. 2021. Integrative taxonomic investigation of *Petaurus breviceps* (Marsupialia: Petauridae) reveals three distinct species. *Zoological Journal of the Linnean Society*, 191(2), pp. 503–527

Cushing, N. and Markwell, K., 2009. Platypus diplomacy: animal gifts in international relations. *Journal of Australian Studies*, 33(3), pp. 255–271

Cuvier, Baron G., 1827. *The Animal Kingdom*. Volume 3. London: G. B. Whittaker

Dampier, W., 1697, pub. 2007. *A New Voyage Round the World*. Warwick, New York: 1500 Books

Dampier, W., 1703, pub. 1961. *A Voyage to New Holland*. Gloucester: Alan Sutton Publishing

Darwin, C., 1839. *Voyages of the Adventure and Beagle, vol. III*. London: Henry Colburn

Darwin, C., 1860. *Letter to Asa Gray (8 June 1860)*, s.l.: Darwin Correspondence Project, 'Letter no. 2825', accessed on 10 April 2020

Darwin, C., 1872a. *Letter to J.D. Hooker (4 August 1872)*,

s.l.: Darwin Correspondence Project, 'Letter no. 8449', accessed on 10 April 2020

Darwin, C., 1872b. *The Origin of Species by Means of Natural Selection, or the Preservation of Favoured Races in the Struggle for Life*. 6th ed. London: John Murray

Decker, O., Eldridge, D. and Gibb, H., 2019. Restoration potential of threatened ecosystem engineers increases with aridity: broad scale effects on soil nutrients and function. *Ecography*, 42, pp. 1370–1382

Department of the Environment and Energy, 2019. *Threatened species strategy – year three progress report*, Canberra: Australian Government

Doherty, T. S., Driscoll, D. A., Nimmo, D. G., Ritchie, E. G. and Spencer, R.-J., 2019. Conservation or politics? Australia's target to kill 2 million cats. *Conservation Letters*, 12, p. e12633

Elder, C., Elder, Sugathapala, S., Akbarian-Tefaghi, L., Langley, J. and Wright, N., 2015. Inter and intra-rater reliability of accuracy of testicular volume evaluation: a simulation study. *Endocrine Abstracts*, 39, EP73

Eldridge, M. D. B. and Browning, T. L., 2002. Molecular genetic analysis of the naturalized Hawaiian population of the brush-tailed rock-wallaby, *Petrogale penicillata* (Marsupialia: Macropodidae). *Journal of Mammalogy*, 83(2), pp. 437–444

Eldridge, M. D. B. and Coulson, G. M., 2015. Family Macropodidae (kangaroos and wallabies). In: D. Wilson & R. Mittermeier, eds. *Handbook of the Mammals of the World. vol. 5. Monotremes and Marsupials*. Barcelona: Lynx Edicions, pp. 630–738

Enjapoori, A. K., Grant, T. R., Nicol, S. C., Lefèvre, C. M., Nicholas, K. R. and Sharp, J. A., 2014. Monotreme lactation protein is highly expressed in monotreme milk and provides antimicrobial protection. *Genome Biology and Evolution*, 6, pp 2754–2773

Evans, G. W., 1822. *A Geographical, Historical, and Topographical Description of Van Diemen's Land*. London: John Souter

Feigin, C. Y., Newton, A. H., Doronina, L., Schmitz, J., Hipsley, C. A., Mitchell, K. J., Gower, G., Llamas, B., Soubrier, J., Heider, T. N., Menzies, B. R., Cooper, A., O'Neill, R. J. and Pask, A. J., 2018. Genome of the Tasmanian tiger provides insights into the evolution and demography of an extinct marsupial carnivore. *Nature Ecology & Evolution*, 2, pp. 182–192

Fender-Barnett, A., 2019. Are Australian snakes really the most dangerous in the world? *CSIROscope*, 27 May. Accessed online 16 May 2020

Fenelon, J. C., McElrea, C., Shaw, G., Evans, A. R., Pyne, M., Johnston, S. D. and Renfree, M., 2021. The unique penile morphology of the short-beaked echidna, *Tachyglossus aculeatus*. *Sexual Development*, advanced publication online DOI: 10.1159/000515145

Field, B., 1825. *Geographical Memoirs on New South Wales*. London: John Murray

Fikes, B. J. and Wilkens, J., 2019. Newly built exhibit will feature two of the odd duck-billed mammals. *The San Diego Union-Tribune*, 21 November. Accessed online 14 April 2020

Finlayson, H. H., 1935. *The Red Centre: Man and Beast in the Heart of Australia*. 1st ed. Sydney: Angus & Robertson Ltd

Flannery, T. F., Rich, T. H., Vickers-Rich, P., Ziegler, T., Veatch. E. G. and Helgen, K. M., 2022a. A review of monotreme (Monotremata) evolution, *Alcheringa: An Australasian Journal of Palaeontology*, 46:1, pp. 3–20

Flannery, T. F., Rich, T. H., Vickers-Rich, P. et al., 2022b. The Gondwanan Origin of Tribosphenida (Mammalia), *Alcheringa: An Australasian Journal of Palaeontology*, 46:3–4, pp. 277–290

Fleay, D., 1947. Platypus land; Australia is a sanctuary for strange creatures that belong to the dim ages. *The New York Times*, 11 May, p. 17

Fleay, D., 1980. *Paradoxical Platypus*. Milton, Queensland: The Jacaranda Press

Fleay-Thomson, R., 2007. *Animals First: The Story of Pioneer Australian Conservationist and Zoologist Dr David Fleay*. Southport, Australia: Keeaira Press

Fox, G. T., 1827. *Synopsis of the Newcastle Museum, late the Allan, formerly the Tunstall, or Wycliffe Museum*. London: T. and J. Hodgson

Freeman, C., 2014. *Paper Tiger: How Pictures Shaped the Thylacine*. Hobart, Tasmania: Forty South Publishing

Friday, A., 2014. Platypus, *Ornithorhynchus anatinus*. *Animal Bytes, a University Museum of Zoology Blog*, 15 May. Accessed online 12 May 2020

Friday, A., 2016. *Francis Maitland Balfour*, Department of Zoology, University of Cambridge website. Accessed online 12 April 2020

Gammage, B., 2011. *The Biggest Estate on Earth*. Crows Nest, NSW: Allen and Unwin

Gelder, K. and Weaver, R., 2020. *The Colonial Kangaroo Hunt*. Melbourne, Victoria: The Miegunyah Press

George, W., 1964. An early European description of an Australasian mammal. *Nature*, 202, pp. 1130–1131

Gould, J., 1863. *The Mammals of Australia*. London: John Gould

Grant, T., 1984. *The Platypus*. 1st ed. Kensington, NSW: New South Wales University Press

Griffiths, M., 1978. *The Biology of the Monotremes*. New York: Academic Press, Inc

Haacke, W., 1885. On the marsupial ovum, the mammary pouch, and the male milk glands of *Echidna hystrix*. *Proceedings of the Royal Society of London*, 38, pp. 72–74

Hall, B. K., 1999. The paradoxical platypus. *BioScience*, 49(3), pp. 211–218

Harman, K., 2018. Explainer: the evidence for the Tasmanian genocide. *The Conversation*, 18 January. Accessed online 12 December 2020

Harris, G., 1808. Description of two new species of *Didelphis* from Van Diemen's Land. *Transactions of the Linnean Society of London*, 9, pp. 174–178

Haygarth, N., 2017. The myth of the dedicated thylacine hunter: stockman–hunter culture and the decline of the thylacine (Tasmanian tiger) in Tasmania during the late nineteenth and early twentieth centuries. *Papers and Proceedings: Tasmanian Historical Research Association*, 64(2), pp. 30–45

Helgen, K. M., Portela Miguez, R., Kohen, J. L. and Helgen, L. E., 2012. Twentieth century occurrence of the long-beaked echidna *Zaglossus bruijnii* in the Kimberley region of Australia. *ZooKeys*, 255, pp. 103–132

Hohnen, R., Ashby, J., Tuft, K. and McGregor, H., 2012. Individual identification of northern quolls (*Dasyurus hallucatus*) using remote cameras. *Australian Mammalogy*, 35, pp. 131–135

Home, E., 1802. A Description of the anatomy of the Ornithorhynchus paradoxus. *Philosophical Transactions of the Royal Society*, 92, pp. 67–84

Hugghins, E. J. and Potter, G. E., 1959. Morphology of the urinogenital system of the opossum (*Didelphis virginiana*). *BIOS*, 30(3), pp. 148–154

Hughes, R. L. and Hall, L. S., 1998. Early development and embryology of the platypus. *Philosophical Transactions of the Royal Society B*, 353(1372), pp. 1101–1114

Hunter, J., 1798a. *Letter to The Literary and Philosophical Society of Newcastle upon Tyne, dated 10th August 1798*. Literary and Philosophical Society Library, Newcastle upon Tyne.

Hunter, J., 1798b. *Letter to Joseph Banks, dated 5th August 1798*. Mitchell Library, State Library of New South Wales, Sydney.

Hurum, J., Luo, Z. and Kielan–Jaworowska, Z., 2006. Were

mammals originally venomous? *Acta Palaeontologica Polonica*, 51(1), pp. 1–11

Indigo, N., Smith, J., Webb, J. and Phillips, B., 2018. Not such silly sausages: Evidence suggests northern quolls exhibit aversion to toads after training with toad sausages. *Austral Ecology*, 43, pp. 592–601

IUCN, 2020. *The IUCN Red List of Threatened Species*, Version 2020–2: IUCN

Jackson, S., 2002. Reproductive behaviour and food consumption associated with the captive breeding of platypus (*Ornithorhynchus anatinus*). *Journal of Zoology*, 256(03), pp. 279–288

Jackson, S., 2015. Family Pseudocheiridae (ring-tailed possums and greater gliders). In: D. Wilson & R. Mittermeier, eds. *Handbook of the Mammals of the World. Vol. 5. Monotremes and Marsupials*. Barcelona: Lynx Edicions, pp. 498–531

Jackson, S. and Vernes, K., 2010. *Kangaroo: a Portrait of an Extraordinary Marsupial*. Crows Nest, NSW: Allen & Unwin

Jamison, J., 1818. Observations on *Ornithorhynchus. Transactions of the Linnean Society of London*, 12(2), pp. 584–585

Janis, C. M., Figueirido, B., DeSantis, L. and Lautenschlager, S., 2020. An eye for a tooth: *Thylacosmilus* was not a marsupial 'saber-tooth predator'. *PeerJ*, 8, e9346

Johnson, C., 2006. *Australia's Mammal Extinctions: A 50,000 Year History*. Cambridge: Cambridge University Press

Johnston, S. D., Smith, B., Pyne, M., Stenzel, D. and Holt, W. V., 2007. One-sided ejaculation of echidna sperm bundles. *The American Naturalist*, 170(6), pp. E162–E164

Jones, K. E., Dickson, B. E., Angielczyk, K. D. and Pierce, S. E., 2021. Adaptive landscapes challenge the 'lateral-to-sagittal' paradigm for mammalian vertebral evolution. *Current Biology*, Volume 31(9), pp. 1883–1892.e7

Klamt, M., Thompson, R. and Davis, J., 2011. Early response of the platypus to climate warming. *Global Change Biology*, 17, pp. 3011–3018

Koh, J. M. S., Bansal, P. S., Torres, A. M. and Kuchel, P. W., 2007. Platypus venom: source of novel compounds. *Australian Journal of Zoology,* 57, pp. 203–210

Lankester, E. R., 1885. Recent progress in biology. *Popular Science Monthly,* 27, pp. 664–668

Lawrence, N., 2011. The Prime Minister and the platypus: A paradox goes to war. *Studies in History and Philosophy of Biological and Biomedical Sciences,* 43(1), pp. 290–297

Lawrence, N., 2017. Churchill's platypus. *BBC Wildlife Magazine,* 27 September, pp. 30–34

Legge, S. et al., 2018. Havens for threatened Australian mammals: the contributions of fenced areas and offshore islands to the protection of mammal species susceptible to introduced predators. *Wildlife Research,* 45(7)

Legge, S. et al., 2020. We need to worry about Bella and Charlie: the impacts of pet cats on Australian wildlife. *Wildlife Research,* 47(8), pp. 523–539

Lewis, J. H., 1996. *Primitive Australian mammals.* In: *Comparative Hemostasis in Vertebrates.* Boston, MA: Springer, pp. 115–122

Long, J., Archer, M., Flannery, T. & Hand, S., 2002. *Prehistoric Mammals of Australia and New Guinea.* Baltimore: The Johns Hopkins University Press

Lord, C. E., 1920. The early history of Bruny Island. *Papers & Proceedings of the Royal Society of Tasmania,* pp. 114–136

Low, T., 2016. *Where Song Began.* US ed. New Haven: Yale University Press

Lydekker, R., 1896. *A Hand-book to the Marsupialia and Monotremata.* London: E. Lloyd

Magondu, B. et al., 2023. Drying dynamics of pellet feces, *Soft Matter,* 19, pp. 723–732

Mahoney, J. A., 1988. Ornithorhynchidae. In: D. Walton, ed. *Zoological Catalogue of Australia, Vol. 5.* Canberra: Australian Government Publishing Service, pp. 7–10

Manger, P. R., Hall, L. S. & Pettigrew, J. D., 1998. The development of the external features of the platypus (*Ornithorhynchus anatinus*). *Philosophical Transactions of the Royal Society B*, 353(1372), pp. 1115–1125

Manger, P. R. & Pettigrew, J. D., 1995. Electroreception and the feeding behaviour of platypus (*Ornithorhynchus anatinus*: Monotremata: Mammalia). *Philosophical Transactions of the Royal Society B*, 347(1322), pp. 359–381

Maynard, D. & Gordon, T., 2014. *Tasmanian Tiger: Precious Little Remains*. Launceston, Tas: The Queen Victoria Museum and Art Gallery

McHugh, E., 2006. *1606: An Epic Adventure*. Sydney: University of New South Wales Press

McKay, H. F., McLeod, P. E., Jones, F. F. & Barber, J. E., 2001. *Gadi Mirrabooka: Australian Aboriginal Tales from the Dreaming*. Santa Barbara, California: Libraries Unlimited

Menzies, B. R., Hildebrandt, T. B. & Renfree, M. B., 2020. Unique reproductive strategy in the swamp wallaby. *Proceedings of the National Academy of Sciences of the United States of America*, 117(11), pp. 5938–5942

Mills, C., 2016. Seed predation paradigm shifts in Australia. *Arid Recovery Blog*, 14 November. Accessed online 25 July 2020

Morrow, G., Andersen, N. A. & Nicol, S. C., 2009. Reproductive strategies of the short-beaked echidna – a review with new data from a long-term study on the Tasmanian subspecies (*Tachyglossus aculeatus setosus*). *Australian Journal of Zoology*, 57(4), pp. 275–282

Moseley, H. N., 1885. On the ova of monotremes. In: *Report of the fifty-fourth meeting of the British Association for the Advancement of Science held at Montreal in August and September 1884*. London: John Murray, p. 777

Moyal, A., 1986. *A Bright and Savage Land: Scientists in Colonial Australia*. Sydney: Collins

Moyal, A., 2001. *Platypus*. Sydney: Allen & Unwin

Naish, D., 2011. Of koalas and marsupial lions: the vombatiform radiation, part I. *Scientific American*, 26 October. Accessed online 11 April 2020

Newton, A. H. et al., 2018. Letting the 'cat' out of the bag: pouch young development of the extinct Tasmanian tiger revealed by X-ray computed tomography. *Royal Society Open Publishing*, Volume 5, p. 171914

Nicol, S., 2015. Family Tachyglossidae (echidnas). In: D. Wilson & R. Mittermeier, eds. *Handbook of the Mammals of the World. Vol. 5. Monotremes and Marsupials*. Barcelona: Lynx Edicions, pp. 34–57

Nicol, S., 2018. Étienne Geoffroy Saint-Hilaire, Richard Owen and monotreme oviparity. *Australian Zoologist*, 40(2), pp. 272–289

Nicol, S. C., 2017. Energy homeostasis in monotreme. *Frontiers in Neuroscience*, 11(195)

Olsen, P., 2010. *Upside Down World: Early European Impressions of Australia's Curious Animals*. Canberra: National Library of Australia

Olsen, P. & Russell, L., 2019. *Australia's First Naturalists: Indigenous Peoples' Contribution to Early Zoology*. Canberra: National Library of Australia

Owen, R., 1832. On the mammary glands of the *Ornithorhynchus paradoxus*. *Philosophical Transactions of the Royal Society of London*, Volume 122, pp. 517–538

Owen, R., 1834a. On the ova of the *Ornithorhynchus paradoxus*. *Philosophical Transactions of the Royal Society of London*, Volume 124, pp. 555–566

Owen, R., 1834b. On the young of the *Ornithorhynchus paradoxus* Blum.. *Proceedings of the Zoological Society of London*, 2(1), pp. 43–44

Owen, R., 1834c. On the generation of the marsupial animals, with a description of the impregnated uterus of the kangaroo. *Philosophical Transactions of the Royal Society of London*, Volume 124, pp. 333–364

Owen, R., 1859. On the fossil mammals of Australia. Part II. Description of a mutilated skull of the large marsupial carnivore (*Thylacoleo carnifex* Owen), from a calcareous conglomerate stratum, eighty miles S. W. of Melbourne, Victoria. *Philosophical Transactions of the Royal Society,* Volume 149, pp. 309–322

Owen, R., 1865. On the marsupial pouches, mammary glands, and mammary foetus of the *Echidna hystrix*. *Philosophical Transactions of the Royal Society of London,* Volume 155, pp. 671–686

Owen, R., 1887. Description of a newly-excluded young of the *Ornithorhynchus paradoxus*. *Proceedings of the Royal Society of London,* Volume 42, p. 391

Paddle, R., 2000. *The Last Tasmanian Tiger.* Cambridge: Cambridge University Press

Parker, W. K., 1885. '*On Mammalian Descent*': the Hunterian Lectures for 1884. London: C. Griffin

Parrish, S. S., 1997. The female opossum and the nature of the New World. *The William and Mary Quarterly,* 54(3), pp. 475–514

Pascoe, B., 2018. *Dark Emu.* London: Scribe

Paterson, A. B., 1933. *The Animals Noah forgot.* Sydney: Endeavour Press

Pian, R., Archer, M. & Hand, S. J., 2013. A new, giant platypus, *Obdurodon tharalkooschild*, sp. nov. (Monotremata, Ornithorhynchidae), from the Riversleigh World Heritage Area, Australia. *Journal of Vertebrate Paleontology,* 33(6), pp. 1255–1259

Pickrell, J., 2018. Climate change implicated in marsupial-lion extinction. *Nature,* 23 October. Accessed online 7 April 2020

Pigott, L. & Jessop, L., 2007. The governor's wombat: early history of an Australian marsupial. *Archives of Natural History,* 34(2), pp. 207–218

Power, J., 2019. Where there's a quill: CSI uses DNA to catch echidna smugglers. *The Sydney Morning Herald*, 16 February. Accessed online 19 April 2020

Pridmore, P. A., 1985. Terrestrial locomotion in monotremes (Mammalia: Monotremata). *Journal of Zoology*, Volume 205, pp. 53–73

Pybus, C., 2020. *Truganini: Journey Through the Apocalypse.* Crows Nest, NSW: Allen & Unwin

Quammen, D., 1996. *The Song of the Dodo.* 1st ed. London: Random House

Radford, J. et al., 2018. Degrees of population-level susceptibility of Australian terrestrial non-volant mammal species to predation by the introduced red fox (*Vulpes vulpes*) and feral cat (*Felis catus*). *Wildlife Research*, 45(7), pp. 645–657

Roberts, R. M., Green, J. A. & Schulz, L. C., 2016. The evolution of the placenta. *Reproduction*, 152(5), pp. R179–R189

Rovinsky, D. S. Evans, A. R. & Adams, J. W. 2021. Functional ecological convergence between the thylacine and small prey-focused canids. *BMC Ecology and Evolution* 21, p. 58

Sample, I., 2018. Scientists unravel secret of cube-shaped wombat faeces. *Guardian*, 18 November. Accessed online 17 May 2020

Samuel, G., 2020. *Independent review of the EPBC Act — interim report,* Canberra: Department of Agriculture, Water and the Environment

San Diego Zoo website. Accessed 14 April 2020. *https:// animals.sandiegozoo.org/animals/platypus*

Science Museum Group Collection Online, 1884. *Caldwell automatic microtome, Cambridge, England. 1959–1964.* Accessed 12 April 2020

Shaw, G., 1792. *The Naturalist's Miscellany, Vol iii.* London: Nodder & Co.

Shaw, G., 1799. *The Naturalist's Miscellany, Vol x.* London: Nodder & Co.

Shaw, G., 1800. *General Zoology.* Vol 1, Pt 1. London: G. Kearsley

Shaw, G., 2006. Reproduction. In: P. Armati, C. Dickman & I. Hume, eds. *Marsupials*. Cambridge: Cambridge University Press, pp. 83–107

Shine, R., Ward-Fear, G. & Brown, G. P. 2020. A famous failure: why were cane toads an ineffective biocontrol in Australia? *Conservation Science and Practice*, 2(12), p. e296

Simons, J., 2008. *Rossetti's Wombat*. London: Middlesex University Press

Simons, J., 2013. *Kangaroo*. London: Reaktion Books Ltd

Sleightholme, S. & Ayliffe, N., 2013. *International thylacine specimen database*. 5th ed

Slezak, M., 2022. Majority of Australia's environment in 'poor' state as Labor blames the Coalition for decade of 'inaction and wilful ignorance', *ABC News*, 18 July 2022. Accessed online 2 February 2023

Smith, D. M., 2020. The return of the platypuses. *The New York Times*, 16 June. Accessed online 18 June 2020

Smith, I., 1972. *The Death of a Wombat*. Melbourne: Wren

Strachey, W., 1612, pub. 1849. *The Historie of Travaile into Virginia Britannia*. London: Hakluyt Society

Strahan, R. & Conder, P., 2007. *Dictionary of Australian and New Guinean Mammals*. Collingwood, Victoria: CSIRO Publishing

Sullivan, H., 2020. Can the world's strangest mammal survive? *The New York Times*, 2020 January. Accessed online 18 June 2020

Summerell, A., Frankham, G., Gunn, P. & Johnson, R., 2019. DNA based method for determining source country of the short beaked echidna (*Tachyglossus aculeatus*) in the illegal wildlife trade. *Forensic Science International*, Volume 295, pp. 46–53

Sutton, P. and Walshe, K., 2021. *Farmers or hunter-gatherers? The Dark Emu debate*. Carlton, Vic: Melbourne University Press

Temminck, C. J., 1824. *Monographie de mammalogie*. Paris: s.n

Tench, W., 1793. *A Complete Account of the Settlement at Port Jackson*. London: G. Nicol and J. Sewell

Thomas, J., Handasyde, K., Parrott, M. L. & Temple-Smith, P., 2018b. The platypus nest: burrow structure and nesting behaviour in captivity. *Australian Journal of Zoology*, 65(6), pp. 347–356

Thomas, J. L., 2018. *Breeding biology of the platypus (*Ornithorhynchus anatinus*)*. PhD thesis: The University of Melbourne

Thomas, J. L., Parrott, M. L., Handasyde, K. A. & Temple-Smith, P., 2018a. Female control of reproductive behaviour in the platypus (*Ornithorhynchus anatinus*), with notes on female competition for mating. *Behaviour*, 155(1), pp. 27–53

Thomas, O., 1888. *Catalogue of the Marsupialia and Monotremata in the Collection of the British Museum (Natural History)*. London: Trustees of the British Museum (Natural History)

Thomson, K., 2008. Jefferson, Buffon and the moose. *American Scientist*, 96(3), p. 200

Travouillon, K. & Phillips, M., 2018. Total evidence analysis of the phylogenetic relationships of bandicoots and bilbies (Marsupialia: Peramelemorphia): reassessment of two species and description of a new species. *Zootaxa*, 4378(2), p. 224

Travouillon, K. et al., 2019. Hidden in plain sight: reassessment of the pig-footed bandicoot, *Chaeropus ecaudatus* (Peramelemorphia, Chaeropodidae), with a description of a new species from central Australia, and use of the fossil record to trace its past distribution. *Zootaxa*, Volume 4566, p. 1

Triggs, B., 1988. *The Wombat*. Kensington, NSW: New South Wales University Press

Tuft, K., 2018. The end of the western barred bandicoot? *Arid Recovery Blog*, 8 March. Accessed online 15 August 2020

Unaipon, D., 1924–1925. *Volume 2: Typescript of 'Legendary Tales of Australian Aborigines'*. Sydney: Mitchell Library, State

Library of New South Wales. Courtesy Ms Judy Kropinyieri

Van Dyck, S. & Strahan, R., 2008. *The Mammals of Australia.* 3rd ed. Sydney: New Holland Publishers

Van Wyhe, J. & Drawhorn, G. M., 2015. 'I am Ali Wallace': The Malay assistant of Alfred Russel Wallace. *Journal of the Malaysian Branch of the Royal Asiatic Society,* 88, Part 1(308), pp. 3–31

Vergnani, L., 2019. Stranger than fiction. *BBC Wildlife*, July, pp. 40–45

Victoria State Government, 2021. *Protecting our iconic platypus.* Melbourne: Victoria State Government

Von Husyett, M., 1994. *The Batavia Journal of Francisco Pelsaert,* Perth: Western Australian Maritime Museum

Wahlquist, C., 2020. Rio Tinto blames 'misunderstanding' for destruction of 46,000-year-old Aboriginal site. *Guardian,* 5 June. Accessed online 6 June 2020

Wallace, A. R., 1869. *The Malay Archipelago.* London: Macmillan and Co

Wallace, A. R., 1876. *The geographical distribution of animals, with a study of the relations of living and extinct faunas as elucidating the past changes of the earth's surface.* London: Macmillan and Co

Walsh, G. P., 1967. Jamison, Sir John (1776–1844). In: *Australian Dictionary of Biography.* National Centre of Biography, Australian National University. Accessed online 13 April 2020

Ward-Fear, G. et al., 2017. Eliciting conditioned taste aversion in lizards: live toxic prey are more effective than scent and taste cues alone. *Integrative Zoology,* 12(2), pp. 112–120

Weisbecker, V. & Beck, R. M. D., 2015. Marsupial and monotreme evolution. In: A. Klieve, L. Hogan, S. Johnston & P. Murray, eds. *Marsupials and Monotremes – Nature's Enigmatic Mammals.* Nova Science Publishers, Inc

Weisbecker, V. & Goswami, A., 2010. Brain size, life history, and metabolism at the marsupial/placental dichotomy.

Proceedings of the National Academy of Sciences of the United States of America, Volume 107, pp. 16216–16221

Weisbecker, V. & Nilsson, M., 2008. Integration, heterochrony, and adaptation in pedal digits of syndactylous marsupials. *BMC Evolutionary Biology*, Volume 8, p. 160

Wellauer, K. & Thomas, K., 2020. WWF report finds 71pc decline in koala numbers across northern NSW bushfire-affected areas. *ABC News*, 2020 September. Accessed online 6 September 2020

Whitley, G. P., 1970. *Early History of Australian Zoology*. Sydney: Royal Zoological Society of New South Wales

Whittington, C. M. & Belov, K., 2007. Platypus venom: a review. *Australian Mammalogy*, Volume 29, pp. 57–62

Woinarski, J. & Burbidge, A., 2016. *Ornithorhynchus anatinus*. *The IUCN Red List of Threatened Species*, e.T40488A-21964009. IUCN. Downloaded 15 August 2020

Woinarski, J., Burbidge, A. & Harrison, P., 2015. Ongoing unraveling of a continental fauna: decline and extinction of Australian mammals since European settlement. *Proceedings of the National Academy of Sciences of the United States of America*, 112(15), pp. 4531–4540

Woodford, J., 2002. *The Secret Life of Wombats*. Melbourne: Text Publishing

Wood, J. G., 1865. *The Illustrated Natural History, Vol I: Mammalia*. London: Routledge, Warne & Routledge

Woodward, H., 1890. *Guide to the Exhibition Galleries of the Department of Geology and Palaeontology in the British Museum (Natural History)*. 6th ed. London: Trustees of the British Museum (Natural History)

Wooller, R. D., 2015. Family Tarsipedidae (honey possum). In: D. Wilson & R. Mittermeier, eds. *Handbook of the Mammals of the World. Vol. 5. Monotremes and Marsupials*. Barcelona: Lynx Edicions, pp. 34–57

Wroe, S. et al., 2013. Climate change frames debate over the extinction of megafauna in Sahul (Pleistocene Australia–New

Guinea). *Proceedings of the National Academy of Sciences of the United States of America,* 110(22), pp. 8777–8781

Wroe, S., McHenry, C. & Thomason, J., 2005. Bite club: comparative bite force in big biting mammals and the prediction of predatory behaviour in fossil taxa. *Proceedings of the Royal Society B,* 272(1563), pp. 619–625

Zhou, Y., Shearwin-Whyatt, L., Li, J. et al., 2021. Platypus and echidna genomes reveal mammalian biology and evolution. *Nature,* 592, pp. 756–762

Zoological Society of London, 1832. *Proceedings of the Committee of Science and Correspondence of the Zoological Society of London. Part II.* London: Richard Taylor

Image Credits

Integrated Images

Colour Plate Section

1 © University of Cambridge. Specimen UMZC A2. 2/33
2 © Public Domain. Image sourced from The Biodiversity Heritage Library
3 © David Fleay Natural History Collection, courtesy of Rosemary Fleay-Thomson
4 © Toby Nowlan
5 © Jack Ashby
6 © University of Cambridge. Specimen UMZC A1. 2/16
7 © Jack Ashby
8 © Jack Ashby
9 incamerastock / Alamy Stock Photo
10 © Jack Ashby
11 © Jack Ashby
12 © Jack Ashby
13 © Jack Ashby
14 © Jack Ashby
15 © Jack Ashby
16 © Jack Ashby
17 © Jack Ashby
18 Courtesy of Smithsonian Institution Archives. Image #SIA_000095_B49_F18_010
19 University of Cambridge / Julieta Sarmiento Photography
20 © UCL Grant Museum of Zoology / Tony Slade. Specimen LDUCZ Z7
21 © Jack Ashby

Index